The Plumbers Handbook

by Joseph P. Almond, Sr.
Revised by Rex Miller

An Audel® Book

Macmillan Publishing Company
New York

Collier Macmillan Canada
Toronto

Maxwell Macmillan International
New York   Oxford   Singapore   Sydney

EIGHTH EDITION

Copyright © 1991 by Macmillan Publishing Company, a division of
Macmillan, Inc.
Copyright © 1969, 1971, 1973, 1976, 1979, 1982, 1985 by Joseph P.
Almond, Sr.

Macmillan Publishing Company
866 Third Avenue, New York, NY 10022

Collier Macmillan Canada, Inc.
1200 Eglinton Avenue East, Suite 200
Don Mills, Ontario M3C 3N1

Production services by the Walsh Group, Yarmouth, ME.

Library of Congress Cataloging-in-Publication Data

Almond, Joseph P.
    The plumbers handbook / by Joseph P. Almond, Sr. — 8th ed./
revised by Rex Miller.
        p.   cm.
    At head of title: Audel.
    Includes index.
    ISBN 0-02-501570-2
    1. Plumbing — Handbooks, manuals, etc.   I. Miller, Rex.
II. Title.
TH6125.A45   1991
696′.1 — dc20   90-40622
                                                        CIP

Macmillan books are available at special discounts for bulk purchases
for sales promotions, premiums, fund-raising, or educational use. For
details, contact:

            Special Sales Director
            Macmillan Publishing Company
            866 Third Avenue
            New York, New York 10022

10 9 8 7 6 5 4 3 2

Printed in the United States of America

Contents

Part II   Plumbing Systems

Part III General Reference Information

Preface

The Plumber's Handbook was designed to assist in the training of apprentices in the plumbing and pipe-fitting trades.

Through the years it has also been a helpful companion on the job to make work easier and more accurate.

This book contains time-saving illustrations plus many tips and shortcuts leading to accurate installations. Such subjects as silver brazing, soft soldering, roughing and repair, lead work, and various pipe fittings and specifications beneficial to fabricators are included.

Also featured are illustrations on vents and venting, a dental office, outside sewage lift station, the Sovent system, septic tanks, water heaters, the Sloan flush valve, solar system water heater, Oasis water cooler, plumbing tools, lots of related math, metric information, and working drawings.

In the classroom or on the job, this book provides simplified and condensed information at your fingerprints.

We would like to thank Dennis Bellville and many friends and associates who have assisted in making this book possible.

JOSEPH P. ALMOND, SR.
REX MILLER

Acknowledgments

As is the case with any book, many people have given generously of their time in assisting the authors. Their efforts, time, and suggestions have been of great value. We would like to take this opportunity to thank each of them for his/her contributions.

Manufacturers of equipment used in the plumbing trade are listed. The following list gives their complete names. These manufacturers and agencies are but a sampling of the many who make the installation and maintenance of plumbing systems possible. In addition, the technical data furnished by these companies and agencies are much appreciated. Without their assistance, this project could have neither appeal nor objectivity.

American National
  Metric Council
Bethesda, Maryland

American Standard
P.O. Box 2003
Brunswick, NJ
(Faucets)

A.O. Smith Corporation
Consumer Products Division
Kankakee, IL 60901
(Hot water heaters)

Cast Iron Soil Pipe Institute
1499 Chain Bridge Road
McLean, VA 22101

Compressed Gas Association,
  Inc.
New York, NY
(Welding)

Copper Development
  Association, Inc.
405 Lexington Avenue
New York, NY 10174

Custom Vacuum Products

Den-Tal-Ez Products

EBCO Manufacturing
   Company
265 N. Hamilton Road
Columbus, OH 43213
(Water coolers)

JET, Inc.
750 Alpha Drive
Cleveland, OH 44143

Lead Industries, Inc.
292 Madison Avenue
New York, NY 10017
(Lead work)

National Bureau of
   Standards
Bureau of Weights &
   Measures
Gaithersburg, MD

NIBCO, Inc.
500 Simpson Avenue
Elkhart, IN 46515
(Silver brazing, soft
   soldering)

Pelton and Crane Co.
P.O. Box 241147
Charlotte, NC 28224
(Dental office equipment)

Plastic Pipe Institute
355 Lexington Avenue
New York, NY 10017

Sloan Valve Co.
10500 Seymour Avenue
Franklin Park, IL 50131
(Sloan flush valve)

Star Dental Products

Syntex Dental Products, Inc.
P.O. Box 896
Valley Forge, PA 19482

U.S. Brass
901 Tenth Street
Plano, TX 75074

Weil Pump Co. and
   Fiberbasin, Inc.
5921 West Dickens
Chicago, IL 60639
(Sewage lift station)

Wrightway Manufacturing
   Company
1050 Central Avenue
Park Forest South, IL 60466

PART I

Techniques, Installation, and Repair

This section includes information that is considered useful, informative, and generally applicable to plumbing today.

1
Plumbing and the Plumber

Plumbing is the art and science of creating and maintaining sanitary conditions in buildings used by humans. It is also the art and science of installing, repairing, and servicing in these same buildings a plumbing system that includes the pipes, fixtures, and appurtenances necessary for bringing in the water supply and removing liquid and water-carried waste.

Many persons have been blinded by getting lime in the eyes either as a powder or as mortar. If you should get lime in your eyes, wash them at once with water. Go to a good doctor for further attention. If lime touches the skin, wash with water, rinse with vinegar, then coat with Vaseline™ in which ordinary kitchen soda bicarbonate has been mixed.

When using electric power tools with abrasive cutting wheels, be sure the tools are properly grounded, particularly if the tool is enclosed in a metal housing or has metal exposed anywhere on the housing. Also, keep eyes and hands properly protected.

When using a ladder, place the ladder base one-fourth of the ladder length away from the structure against which the ladder is leaning.

The maximum height of horses supporting scaffold platforms is 16 feet (4.88 m).

In trenches in hard, compact ground, vertical braces should be located every 8 feet (2.44 m) and horizontal stringers every 4 feet (1.22 m).

Safety — A good item to have on any job requiring the use of an open flame, such as a Prestolite™ torch, melting pot tank, or welding equipment, is an approved type CO_2 or a dry powder fire extinguisher. *Note:* A carbon tetrachloride extinguisher at one time was considered the best agent to use, but it was found to release a deadly gas. It can be especially dangerous if used in a confined place.

The dry powder CO_2 extinguisher, which is actually composed of 99% baking soda and 1% drying agent, is also one of the best agents to use when fighting fires in or around electrical equipment.

A vacuum relief valve should be installed on a copper tank to prevent collapse in the event of a vacuum occurrence.

Seventy-five feet (23 meters) is the maximum spacing for rainwater leaders.

One hundred and fifty square feet (14 square meters) of roof area to 1 square inch (6.5 square centimeters) of leader area.

The instrument for measuring relative humidity is called a hygrometer.

Transfer of heat occurs in three ways:
1. Convection
2. Radiation
3. Conduction

Bread can be packed in a water line to hold back water long enough to solder a joint where water continues to trickle, even if valve or valves are shut off.

To replace a defective section or add a fitting to a rigid pipeline, in most cases you will need to buckle the new portion back into the line in sections.

For cast iron pipe with lead joints and hard copper tubing, three sections and four joints will be necessary.

On plastic pipe through 2″ (51 mm), two section 4′ (122 cm) long and three joints will generally work. See example of 2″ (51 mm) plastic PVC pressure water line below, after it was repaired and reassembled.

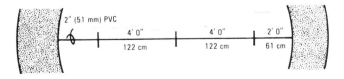

Fig. 1-1.

Suction pumps, barometers, and siphons depend on the natural pressure of the atmosphere in order to function.

A *vacuum* is a space entirely devoid of matter. A *partial vacuum* is a space where an air pressure exists that is less than atmospheric.

2
Plumbing Tools

Basic Tools

Thanks are due the Ridge Tool Company, 400 Clark Street, Elyria, OH 44035, manufacturer of Ridgid™ pretested work-saver tools, for providing the tool illustrations in this section.

Fig. 2-1. Ballpeen hammer.

Fig. 2-2. Pipe wrench. Four sizes: 2″ through 8″ (51-203 mm).

Here is a wrench that lets one person do the work of two. A short handle makes it easy to access frozen joints even in tight quarters.

Fig. 2-3. RIDGID straight pipe wrenches. Ten sizes, 6″ through 60″ (152 through 1524 mm).

These, the world's most popular pipe wrenches, are known for the brutal punishment they can take because of their extra built-in toughness. Before shipment, every wrench is work tested. The housing is replaced free if it ever breaks or distorts. Replaceable jaws are made of hardened alloy steel. A full-floating hook jaw assures instant grip and easy release. Spring suspension eliminates the chance that jaws could jam or lock on pipe. A handy pipe scale and large, easy-to-spin adjusting nut give fast, one-hand setting to pipe size, and a comfort-grip, malleable iron I-beam handle has a convenient hang-up hole.

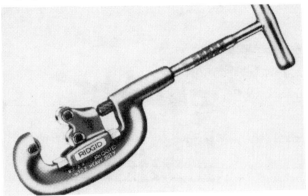

Fig. 2-4. Heavy-duty pipe cutter; cuts pipe ⅛″ through 2″ (3 through 51 mm).

Lever-type tube benders come in six sizes from ³/₁₆″ through ½″ (5 through 13 mm) O.D. They make fast, accurate bends on soft and hard copper, brass, aluminum, steel, and stainless steel tube. The form-handle has "gain marks" for accurate tube measurements before cutting. Pieces can be

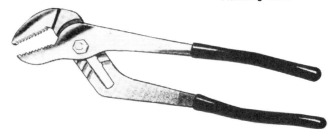

Fig. 2-5. Channel-lock pliers.

Fig. 2-6. Swaging tools.

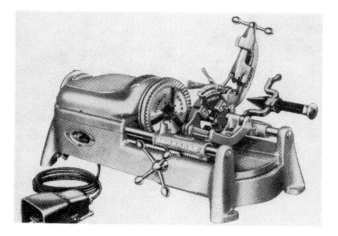

Fig. 2-7. Pipe and bolt threading machine for ⅛″ through 2″ (3 through 51 mm) pipe. This machine cuts, threads, reams, and oils.

pre-cut to proper length, eliminating extra cutting and wasted material. The handles are wide apart when completing a full 180° bend, thus no knuckle cracking.

The K-37 drain cleaner (Fig. 2-24) speed cleans ¾″ (19 mm) through 3″ (76 mm) lines without removing trap or cross bars. This rugged, compact unit with its unique, dual-action clutch represents the latest in drain gun design. A slide action hand grip permits operation "on the fly" while the drum rotates. There is no need to stop the unit to advance or retract the cable. The knurled spin chuck is used for tough obstructions. A positive clutch lock transfers maximum torque to the stoppage and absorbs contact shock before it gets into the drum. Drum capacity: 35′ (10.6 m) with ⅜″ (9.5 mm) cable; 50′ (15 m) with 5/16″ (8 mm) cable.

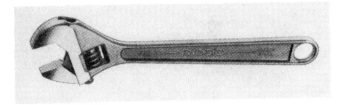

Fig. 2-8. Adjustable wrench, often called crescent wrench.

Fig. 2-9. **RIDGID NO. 342 Internal Wrench.** Holds closet spuds and bath, basin, and sink strainers through 2″ (51 mm). Also handy when installing or extracting 1″ through 2″ (25 through 51 mm) nipples without damage to threads.

Fig. 2-10. **RIDGID Basin Wrenches. (A) RIDGID No. 1010 Basin Wrench** has solid 10″ (254 mm) shank. **(B) RIDGID Nos. 1017 and 1019 Basin Wrenches** have telescopic shanks for 4 lengths from 10″ through 17″ (254 through 431 mm).

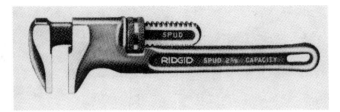

Fig. 2-11. Spud wrench; capacity 2⅝″ (66.5 mm).

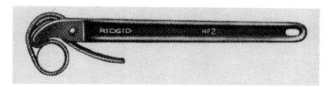

Fig. 2-12. Strap wrench — ⅛″ through 2″ (3 through 51 mm).

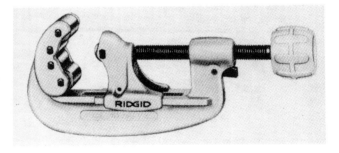

Fig. 2-13. Quick-acting tubing cutter — ¼″ through 2⅝″ (6 through 66.5 mm).

Fig. 2-14. Closet auger, used for water closet and urinal stoppage.

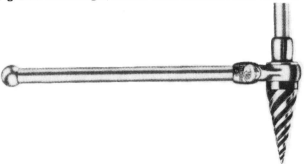

Fig. 2-15. Spiral ratchet pipe reamer — ⅛″ through 2″ (3 through 51 mm).

Fig. 2-16. Heavy-duty 2″ (51 mm) chain wrench.

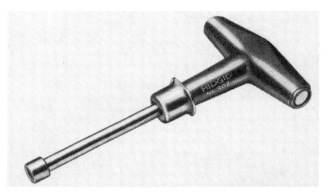

Fig. 2-17. Torque wrench for cast iron and no-hub soil pipe. Pre-set for 60 inch lbs. (67.9 Nm) of torque.

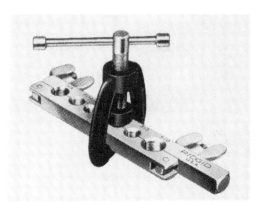

Fig. 2-18. Flaring tool — will flare tubing size from 3/16″ (4.8 mm), 1/4″ (6.4 mm), 5/16″ (7.9 mm), 3/8″ (9.5 mm), 7/16″ (11.1 mm), 1/2″ (12.7 mm), and 5/8″ (15.9 mm).

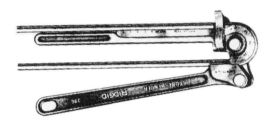

Fig. 2-19. Lever-type tube benders.

Fig. 2-20. Straight snips.

Miscellaneous Tips on Tools

When a soldering iron is overheated, the bright areas will show bluish tarnish, and the tinning on the bit will be dark, dull, and powdery in appearance.

Fig. 2-21. Ratchet cutter for cast iron pipe, 2″ through 6″ (51 through 152 mm).

Fig. 2-22. Soil pipe assembly tool for 2″ through 8″ (51 through 203 mm) pipe.

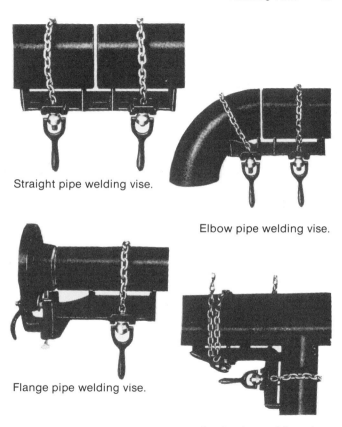

Straight pipe welding vise.

Elbow pipe welding vise.

Flange pipe welding vise.

Angle pipe welding vise.

Fig. 2-23. Pipe welding vises.

Fig. 2-24. K-37 drain cleaner.

A friction clamp for use with brass pipe in a regular pipe vise is made by cutting a pipe coupling in half lengthwise, then lining it with sheet metal.

Use light machine oil when oiling a rule.

Hacksaw blade manufacturers recommend that a blade with 24 teeth per inch (10 teeth per cm) be used for cutting angle iron or pipe; to cut hanger rod, 18 teeth per inch (7 teeth per cm) is satisfactory. For light-gage band iron and thin-wall tubing, 32 teeth per inch (13 teeth per cm) is best.

The size of a pipe wrench is measured from the inside top of the movable jaw to the end of the handle with the wrench fully opened.

The wrench size for a flange bolt is: bolt size $\times 2 + \frac{1}{8}''$; millimeters $\times 2 + 3$ mm.

Acquire the habit of using two wrenches when tightening or loosening pipe — you will avoid many unnecessary problems.

3
Tips for the Beginning Plumber

Here is a variety of useful information — facts a plumber must know — gleaned from years of plumbing experience. The apprentice or even journeyman should find these tips useful and helpful.

Reading Blueprints

When a measurement is taken from a blueprint it should be checked from both ends of the building to ensure accuracy.

✳ *Datum* is an established level or elevation from which vertical measurements are taken; a *bench mark* (BM) is a measure on which all other elevations are based.

All buildings have a *base elevation* from which all other elevations and grades are determined; some plans use 100.0', others use 0.00'.

Example: A basement floor level 91.5', using 100.0' as the base level, would indicate a basement floor level 8'6" below the first-floor level.

Bench marks permit the plumber to locate the elevations pertaining to his work. A 96' or a 104' bench mark would indicate 4 feet below or 4 feet above *finish floor* (FF). Examples of bench marks: ·

$$\text{BM} \qquad \text{F.F.} + 4.0'$$
$$\text{F.F.} - 2.0'$$

A *sectional elevation drawing* would provide the plumber with information as to width and height of a specific portion of the structure.

Elevation measurements on piping plans are called *invert elevations.*

A figure in isometric position lies with one corner directly in front of you. The back corner is tilted to a 30° angle.

A building plan may denote an invert elevation of 0.325' at one end of a pipeline and 0.400' at the opposite end — a difference of 0.75'. By multiplying 0.75' × 12, you will find that the difference in inches between the two points will be 9 inches (Table 3-1).

Shooting Grade Levels

Plumbers are often called on to set grade levels for various piping elevations, including catch basins, floor drains, and many other grade levels associated with their work. It therefore behooves every plumber and fitter to become familiar with this very important phase of the piping industry.

Two main parts to shooting grade levels are: (1) the dumpy level and (2) the leveling rod.

The dumpy level (named after its inventor) is a surveyor's level with a short inverted telescope rigidly affixed, and rotating in a horizontal plane only. This level mounts onto a tripod.

The leveling rod is a graduated rod used in measuring the vertical distance between a point on the ground and the line of sight of a surveyor's level, or dumpy level. This rod is marked off in 10ths and 100ths, and its scale is known as the engineer's scale.

Each foot on the leveling rod is divided into 10ths and each 10th is divided into 10ths: thus, 100 marks in all. Every 10 marks there is a number from 1 to 9; following each 9 will be the proper foot mark.

Always remember that each foot on the leveling rod equals 12 inches; each tenth of this foot equals 1.2 inches; each one (1) of the 100 marks contained in the engineer's foot equals 0.01, or one-one-hundredth of a foot. The exception would be the leveling rods marked off in two-one-hundredths (0.02) of a foot. They have 5 marks to each tenth, 50 marks to each foot.

To determine inches from hundredths of a foot you simply multiply by 12.

Example:

$$
\begin{array}{r}
0.54' \\
\times\ 12 \\
\hline
108 \\
54 \\
\hline
6.48 \text{ rounded to } 6\frac{1}{2}''
\end{array}
$$

Refer to Fig. 3-1, which shows a one (1) foot section of a leveling rod marked off in 0.01 of a foot.

Note the arrow and where it is pointing; as you will see, it points to 0.54' — $^{54}/_{100}$ of a foot, thus: 0.54 × 12 = 6½ inches rounded.

Table 3-1. Converting Inches to Decimal Parts of a Foot

1″ = 0.083′	7″ = 0.5833′
2″ = 0.1666′	8″ = 0.6667′
3″ = 0.25′	9″ = 0.75′
4″ = 0.333′	10″ = 0.8333′
5″ = 0.4167′	11″ = 0.9333′
6″ = 0.50′	12″ = 1.00′

To convert decimal feet to inches, multiply by 12: you may then change decimal inches to inches and fractions.

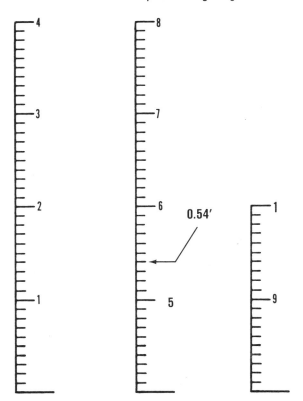

Fig. 3-1. Lower section of a leveling rod.

Heating Systems

In heating systems the compression tank plays an important part in the economical operation of the system. When water in the system is heated it expands. If no tank were installed, the expanding water would be forced out through the relief valve. In that case, cool water would be drawn in to replace the water lost by expansion.

Extra fuel is used to heat this cold water; also, the constant adding of water brings in foreign matter, such as sediment or lime. This results in scaling of the boiler with an ever increasing amount of fuel required for heating.

Note: Water in a heating system, heated from 32°F to 212°F (0°C to 100°C) will expand approximately $\frac{1}{23}$ of its original volume.

Convection is the method used for transferring heat in a gravity domestic hot water circulation system. Convection, or circulating currents, is produced due to the difference in weight of water at different temperatures.

Absolute pressure is gage pressure plus atmospheric pressure.

Water Heaters

The maximum acceptable temperature for domestic hot water is from 140°F to 160°F (60°C to 71°C). Use of automatic laundry and dishwashing machines makes 160°F (71°C) preferable. Temperatures above 160°F (71°C) are not recommended, as they cause increased corrosion, increased deposit of lime, waste of fuel, more rapid heat loss by radiation, danger of scalding, and other accidents.

If a 30-gallon (113.56-liter) hot water boiler is insulated with a tank jacket, 30% of the total amount of fuel usually burned can be saved.

If hot water pipes are insulated, the heat loss from pipes is reduced by up to 80%.

There should be a minimum of 6″ (15 cm) between an uninsulated water heater and any unprotected wood.

In some water heaters a special device is used to prevent corrosion; this device is called a *protector rod*.

Water Supply

Storage tanks up to 82 gallons (310.4 liter) capacity are tapped for 1″ (25 mm) connections; tanks over that size are tapped a minimum of 1¼″ (32 mm) generally.

The dip tube on a cold water supply should terminate 8″ (20 cm) above the bottom of the tank.

The standard length of asbestos cement water main pressure pipe is 13 feet (approximately 4 meters).

When water solidifies it becomes lighter.

The amount of heat required to change ice to liquid water is 144 BTUs per pound (335 joules per kilogram).

The installation of a water softener in a residential piping system causes fairly high pressure loss.

Fixtures, Valves, and Fittings

When ordering piping elbows, an example would be: 6 each 1¼″ (32 mm) copper, PVC, or galvanized 90° ells. If reducing elbows are ordered, list the largest measurement first: 6 each 1¼″ × 1″ (32 × 25 mm) copper 90° reducing ells.

When ordering tees, you would begin by listing the largest measurement on the run or flow line — always listing the line measurement last. *Example:* 6 each 1¼″ × 1″ × 1½″ (32 × 25 × 38 mm) copper tees; 1½″ is the branch line measurement.

In plumbing, the pipe size measurement given is always N.P.S. — I.D. or inside diameter. In air conditioning and refrigeration, pipe and tubing are called and ordered by their O.D. or outside diameter measurement. Therefore, a ¼″ (19 mm) copper pipe in plumbing would be called ⅞″ (22 mm) pipe or tubing in refrigeration.

Flare fittings are sold and ordered by their O.D. measurement.

Brass fittings contain 85% copper, 5% zinc, 5% tin, and 5% lead.

Approximate heights above F.F. or floor level rims for plumbing fixtures are: sink — 36″ (91 cm), built-in bathtub — 16″ (41 cm), water closet — 15″ (38 cm), lavatory — 31″ (79 cm), wash or laundry tray — 34″ (86 cm).

An air gap of 1″ (2.5 cm) to 2″ (5 cm) between the flood level rim of a fixture and the water supply opening is considered safe.

A vacuum breaker should be at least 6″ (15 cm) above the flood level rim or 6″ (15 cm) above the top of the unit.

Globe valves have a machined seat and a composition disc and usually shut off tight, while gate valves may leak slightly when closed, particularly if frequently operated, due to wear between the brass gates and the faces against which they operate. Globe valves create more flow resistance than gate valves.

When copper and steel come in contact with each other, especially when dampness is present, a chemical action called *electrolysis* is created.

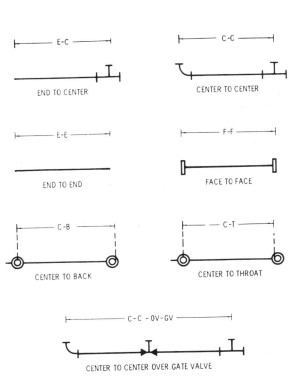

Fig. 3-2. Common terms to remember.

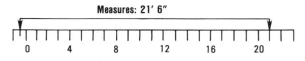

Measures: 21' 6"

0   4   8   12   16   20

Scale: ⅛" = 1 Foot

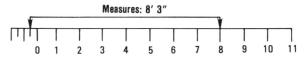

Measures: 8' 3"

0 1 2 3 4 5 6 7 8 9 10 11

Scale: ¼" = 1 Foot

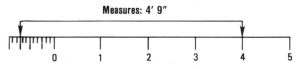

Measures: 4' 9"

0   1   2   3   4   5

Scale: ½" = 1 Foot

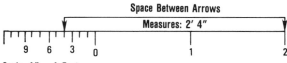

Space Between Arrows
Measures: 2' 4"

9 6 3   0   1   2

Scale: 1" = 1 Foot

Fig. 3-3. Scale-rule examples for study.

Table 3-2. Millimeters to Decimals

mm	Deci-mal	mm	Deci-mal	mm	Deci-mal	mm	Deci-mal	mm	Deci-mal
0.01	.00039	0.41	.01614	0.81	.03189	21	.82677	61	2.40157
0.02	.00079	0.42	.01654	0.82	.03228	22	.86614	62	2.44094
0.03	.00118	0.43	.01693	0.83	.03268	23	.90551	63	2.48031
0.04	.00157	0.44	.01732	0.84	.03307	24	.94488	64	2.51969
0.05	.00197	0.45	.01772	0.85	.03346	25	.98425	65	2.55906
0.06	.00236	0.46	.01811	0.86	.03386	26	1.02362	66	2.59843
0.07	.00276	0.47	.01850	0.87	.03425	27	1.06299	67	2.63780
0.08	.00315	0.48	.01890	0.88	.03465	28	1.10236	68	2.67717
0.09	.00354	0.49	.01929	0.89	.03504	29	1.14173	69	2.71654
0.10	.00394	0.50	.01969	0.90	.03543	30	1.18110	70	2.75591
0.11	.00433	0.51	.02008	0.91	.03583	31	1.22047	71	2.79528
0.12	.00472	0.52	.02047	0.92	.03622	32	1.25984	72	2.83465
0.13	.00512	0.53	.02087	0.93	.03661	33	1.29921	73	2.87402
0.14	.00551	0.54	.02126	0.94	.03701	34	1.33858	74	2.91339
0.15	.00591	0.55	.02165	0.95	.03740	35	1.37795	75	2.95276
0.16	.00630	0.56	.02205	0.96	.03780	36	1.41732	76	2.99213
0.17	.00669	0.57	.02244	0.97	.03819	37	1.45669	77	3.03150
0.18	.00709	0.58	.02283	0.98	.03858	38	1.49606	78	3.07087
0.19	.00748	0.59	.02323	0.99	.03898	39	1.53543	79	3.11024
0.20	.00787	0.60	.02362	1.00	.03937	40	1.57480	80	3.14961
0.21	.00827	0.61	.02402	1	.03937	41	1.61417	81	3.18898
0.22	.00866	0.62	.02441	2	.07874	42	1.65354	82	3.22835
0.23	.00906	0.63	.02480	3	.11811	43	1.69291	83	3.26772
0.24	.00945	0.64	.02520	4	.15748	44	1.73228	84	3.30709
0.25	.00984	0.65	.02559	5	.19685	45	1.77165	85	3.34646
0.26	.01024	0.66	.02598	6	.23622	46	1.81102	86	3.38583
0.27	.01063	0.67	.02638	7	.27559	47	1.85039	87	3.42520
0.28	.01102	0.68	.02677	8	.31496	48	1.88976	88	3.46457
0.29	.01142	0.69	.02717	9	.35433	49	1.92913	89	3.50394
0.30	.01181	0.70	.02756	10	.39370	50	1.96850	90	3.54331
0.31	.01220	0.71	.02795	11	.43307	51	2.00787	91	3.58268
0.32	.01260	0.72	.02835	12	.47244	52	2.04724	92	3.62205
0.33	.01299	0.73	.02874	13	.51181	53	2.08661	93	3.66142
0.34	.01339	0.74	.02913	14	.55118	54	2.12598	94	3.70079
0.35	.01378	0.75	.02953	15	.59055	55	2.16535	95	3.74016
0.36	.01417	0.76	.02992	16	.62992	56	2.20472	96	3.77953
0.37	.01457	0.77	.03032	17	.66929	57	2.24409	97	3.81890
0.38	.01496	0.78	.03071	18	.70866	58	2.28346	98	3.85827
0.39	.01535	0.79	.03110	19	.74803	59	2.32283	99	3.89764
0.40	.01575	0.80	.03150	20	.78740	60	2.36220	100	3.93701

Drains and Sewers

The *building drain* is the lowest horizontal piping inside the building. It connects with the building sewer.

The *building sewer* extends from the main sewer or other disposal terminal to the building drain at a distance of approximately 5′ (152 cm) from the foundation wall.

Public sewer manholes can be used to verify main sewer elevations and direction of flow.

Gases found in sewer air are carbon monoxide, methane, hydrogen sulphide, carbon dioxide, gasoline, ammonia, sulphur dioxide, illuminating gas.

Primary treatment in a sewage treatment plant removes floating and settleable solids. *Secondary treatment* removes dissolved solids.

Notes: Relief line from relief valve is generally piped to the outside, 12″ (305 mm) or less above ground level, elbow and nipple turned down. The nipple should not be threaded on outlet end.

The dip tube on cold water supply should terminate 8″ (20 cm) above bottom of tank.

In some water heaters a special device is used to prevent corrosion. This device is called a protector rod.

The temperature-sensitive element of the relief valve should be installed directly in the tank proper so that it is in direct contact with the hot water.

The recommended and safest procedure is to place the relief valve in a separate tapping either at the top of the tank or within four inches from the top if tapping is located at the side.

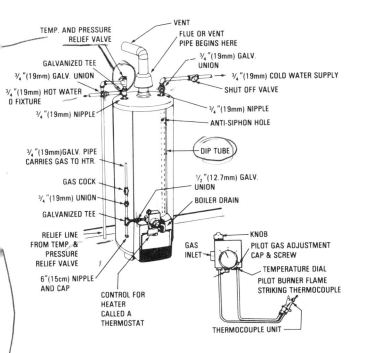

Fig. 3-4. Gas hot water heater in average residence.

Gas Water Heaters

If there are no separate tappings on the water heater, then place the relief valve as shown; however, use nipples as short as possible.

Never install a check valve in the water supply to a water heater, because it would confine pressure in the tank and result in an accident if the relief valve did not operate.

There is a small hole drilled in the dip tube near the top; this hole admits air to the cold water piping to break siphonic action.

The nipple and cap at the bottom of the tee where the gas supply turns into the heater form a dirt and drip pocket.

Procedure for Lighting Heater

1. Turn gas cock on control to "off" position, and dial assembly to lowest temperature position.
2. Wait approximately 5 minutes to allow gas that may have accumulated in burner compartment to escape.
3. Turn gas cock handle on control to "pilot" position.
4. Fully depress set button and light pilot burner.
5. Allow pilot to burn approximately 1 minute before releasing set button. If pilot does not remain lighted, repeat operation.
6. Turn gas cock handle on control to "on" position and turn dial assembly to desired position. The main burner will then ignite.

Note: Adjust pilot burner air shutter (if provided) to obtain a soft blue flame.

4
Dental Office

Dental Office Piping Illustrations

The following drawings, illustrations, and related information are intended to serve as a guide and/or example of a typical
dental chair installation.

The Den-Tal-Ez®, Model CMU-D chair-mounted delivery
system is shown in Fig. 4-1. Fig. 4-2 shows a Hustler II® compressor. Fig. 4-3 shows a deaquavator in connection with this
compressor. Fig. 4-4 gives a plan view of a dental chair layout,
showing minimum location from walls and plumbing fixtures.
Fig. 4-5 illustrates air piping from the compressor through the
deaquavator to the dental chairs. Fig. 4-6 illustrates the piping
arrangement from the evacuator to the dental chairs. Fig. 4-7
features the template used to position the supply piping to the
dental chair utility box. Fig. 4-8 is an elevation view of the utility
layout, and Fig. 4-9 displays the kit items. Fig. 4-10 shows a
view of the utility locations. Fig. 4-11 illustrates the connections
in the evacuation system shown in Fig. 4-6.

Dental Office Notes

Before proceeding with the plumbing or piping installation,
consult all applicable utility codes and regulations.

Central vacuum, gravity drain, air supply, and water supply
piping through the floor should be ½″ or 13 mm I.D. rigid
copper tubing (see Tables 4-2 and 4-3 for fractions to decimal to
mm).

Refer to Fig. 4-6. No part of the exhaust line should be more
than 3′ or 91 cm above the level of the waste connection

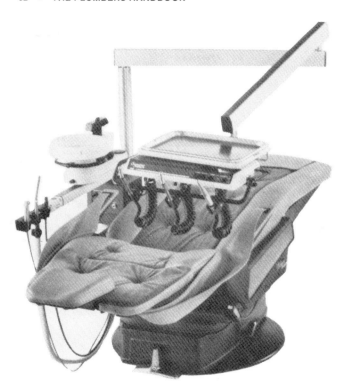

Fig. 4-1. Den-Tal-Ez chair.

on the vacuum pump. Also, there should be a 1″ slope per 20 feet, or 25 mm every 6 meters toward the vacuum producer. The evacuator can be located wherever it is convenient, or close to the compressor.

Fig. 4-2. Hustler II Compressor.

On the air line from the compressor to the utility, rigid copper tubing — type "L" with brazed joints — is recommended. The air supply should be 80 to 100 psi or 552 to 689 kPa.

On the drain from the floor, you should provide a 1½" or 2", 4 or 5 cm P-trap below the floor, reducing the pipe as it goes through the floor.

The water supply pressure should be 40 to 45 psi, or 276 to 310 kPa.

Table 4-1. Decimals to Millimeters

Decimal	mm	Decimal	mm	Decimal	mm	Decimal	mm	Decimal	mm
0.001	0.0254	0.140	3.5560	0.360	9.1440	0.580	14.7320	0.800	20.3200
0.002	0.0508	0.150	3.8100	0.370	9.3980	0.590	14.9860	0.810	20.5740
0.003	0.0762	0.160	4.0640	0.380	9.6520			0.820	20.8280
0.004	0.1016	0.170	4.3180	0.390	9.9060	0.600	15.2400	0.830	21.0820
0.005	0.1270	0.180	4.5720			0.610	15.4940	0.840	21.3360
0.006	0.1524	0.190	4.8260	0.400	10.1600	0.620	15.7480	0.850	21.5900
0.007	0.1778			0.410	10.4140	0.630	16.0020	0.860	21.8440
0.008	0.2032	0.200	5.0800	0.420	10.6680	0.640	16.2560	0.870	22.0980
0.009	0.2286	0.210	5.3340	0.430	10.9220	0.650	16.5100	0.880	22.3520
		0.220	5.5880	0.440	11.1760	0.660	16.7640	0.890	22.6060
0.010	0.2540	0.230	5.8420	0.450	11.4300	0.670	17.0180		
0.020	0.5080	0.240	6.0960	0.460	11.6840	0.680	17.2720	0.900	22.8600
0.030	0.7620	0.250	6.3500	0.470	11.9380	0.690	17.5260	0.910	23.1140
0.040	1.0160	0.260	6.6040	0.480	12.1920			0.920	23.3680
0.050	1.2700	0.270	6.8580	0.490	12.4460	0.700	17.7800	0.930	23.6220
0.060	1.5240	0.280	7.1120			0.710	18.0340	0.940	23.8760
0.070	1.7780	0.290	7.3660	0.500	12.7000	0.720	18.2880	0.950	24.1300
0.080	2.0320			0.510	12.9540	0.730	18.5420	0.960	24.3840
0.090	2.2860	0.300	7.6200	0.520	13.2080	0.740	18.7960	0.970	24.6380
		0.310	7.8740	0.530	13.4620	0.750	19.0500	0.980	24.8920
0.100	2.5400	0.320	8.1280	0.540	13.7160	0.760	19.3040	0.990	25.1460
0.110	2.7940	0.330	8.3820	0.550	13.9700	0.770	19.5580	1.000	25.4000
0.120	3.0480	0.340	8.6360	0.560	14.2240	0.780	19.8120		
0.130	3.3020	0.350	8.8900	0.570	14.4780	0.790	20.0660		

100 mm = 1 cm
1000 mm = 1 meter
1″ = 25.4 mm

milli means $\frac{1}{1000}$ or 0.001
centi means $\frac{1}{100}$ or 0.01

Fig. 4-2 shows a Hustler II compressor. It is one of the largest air power plants specifically designed for the dental profession. Its 30-gallon (113.6-liter) tank and powerful 12.2 scfm makes it ideal for multiple operatories, clinics, or group practices. This heavy duty compressor operates with two 1½ hp motors and pumps working simultaneously. The Hustler II is equipped with separate on/off switches for each compressor.

Table 4-2. Fractions to Decimals to Millimeters

Fraction	Decimal	mm	Fraction	Decimal	mm
1/64	0.0156	0.3969	33/64	0.5156	13.0969
1/32	0.0312	0.7938	17/32	0.5312	13.4938
3/64	0.0469	1.1906	35/64	0.5469	13.8906
1/16	0.0625	1.5875	9/16	0.5625	14.2875
5/64	0.0781	1.9844	37/64	0.5781	14.6844
3/32	0.0938	2.3812	19/32	0.5938	15.0812
7/64	0.1094	2.7781	39/64	0.6094	15.4781
1/8	0.1250	3.1750	5/8	0.6250	15.8750
9/64	0.1406	3.5719	41/64	0.6406	16.2719
5/32	0.1562	3.9688	21/32	0.6562	16.6688
11/64	0.1719	4.3656	43/64	0.6719	17.0656
3/16	0.1875	4.7625	11/16	0.6875	17.4625
13/64	0.2031	5.1594	45/64	0.7031	17.8594
7/32	0.2188	5.5562	23/32	0.7188	18.2562
15/64	0.2344	5.9531	47/64	0.7344	18.6531
1/4	0.2500	6.3500	3/4	0.7500	19.0500
17/64	0.2656	6.7469	49/64	0.7656	19.4469
9/32	0.2812	7.1438	25/32	0.7812	19.8438
19/64	0.2969	7.5406	51/64	0.7969	20.2406
5/16	0.3125	7.9375	13/16	0.8125	20.6375
21/64	0.3281	8.3344	53/64	0.8281	21.0344
11/32	0.3438	8.7312	27/32	0.8438	21.4312
23/64	0.3594	9.1281	55/64	0.8594	21.8281
3/8	0.3750	9.5250	7/8	0.8750	22.2250
25/64	0.3906	9.9219	57/64	0.8906	22.6219
13/32	0.4062	10.3188	29/32	0.9062	23.0188
27/64	0.4219	10.7156	59/64	0.9219	23.4156
7/16	0.4375	11.1125	15/16	0.9375	23.8125
29/64	0.4531	11.5094	61/64	0.9531	24.2094
15/32	0.4688	11.9062	31/32	0.9688	24.6062
31/64	0.4844	12.3031	63/64	0.9844	25.0031
1/2	0.5000	12.7000	1	1.0000	25.4000

Fig. 4-3. Deaquavator.

All Hustler models are equipped with an automatic tank drain that automatically removes condensate at the end of each pumping cycle.

The Hustler II is 45″ (114 cm) wide, 32″ (81 cm) high, and 19″ (48 cm) deep.

The deaquavator shown in Figure 4-3 provides desert-dry air by removing more than 90% of all humidity and oil vapors from up to ten cubic feet, or 283,168 cubic centimeters (cm^3) of air per minute. This very dry air minimizes the most frequent cause of costly handpiece repair. The deaquavator operates on the same principle as a household refrigerator. It provides trouble-free performance for long periods.

The deaquavator is 18″ (457 mm) wide, 14″ (356 mm) high, and 13″ (330 mm) deep. It weighs 62 pounds (28 kg). It operates on a 120 volt, 60 Hz, single-phase power source.

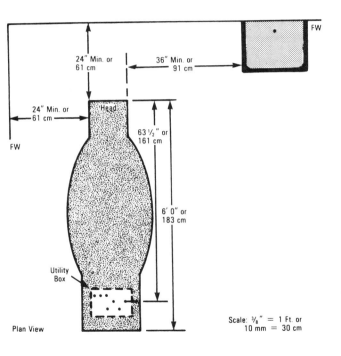

24" Min. or
61 cm

36" Min. or
91 cm

FW

24" Min. or
61 cm

Head

63 ½" or
161 cm

6' 0" or
183 cm

FW

Utility
Box

Plan View

Scale: ⅜" = 1 Ft. or
10 mm = 30 cm

Fig. 4-4. Dental chair layout.

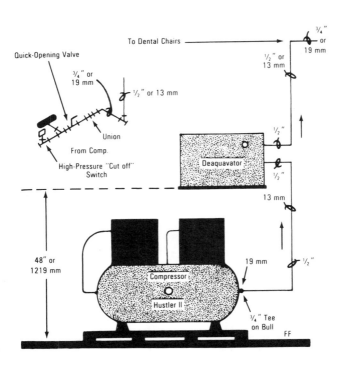

Fig. 4-5. Air piping to dental chairs.

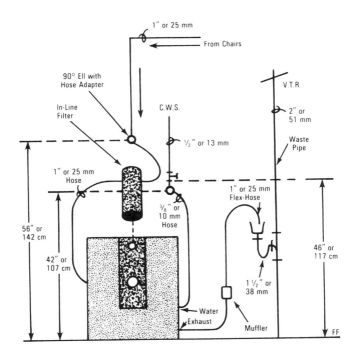

Fig. 4-6. Dynamic-Dual Evacuation System.

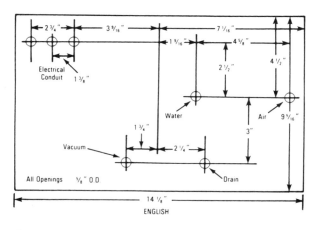

ENGLISH

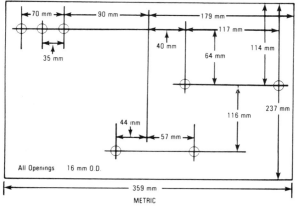

METRIC

Fig. 4-7. Template of USC III.

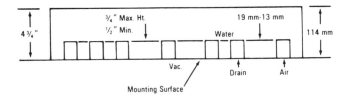

Fig. 4-8. Elevation view of utility layout.

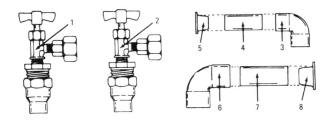

UTILITY STOP KIT NO. 3552-010

1. Water Stop
2. Air Stop
3. Copper Elbow Vacuum
4. Tubing Adaptor for No. 3
5. Plug for Vacuum
6. Copper Elbow Gravity Drain
7. Tubing Adaptor for No. 6
8. Plug for Gravity Drain

Fig. 4-9. Utility Stop Kit items.

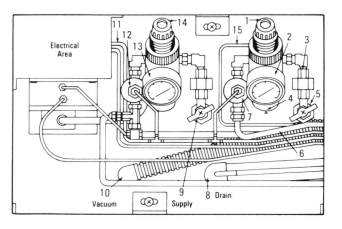

Fig. 4-10. Utility locations.

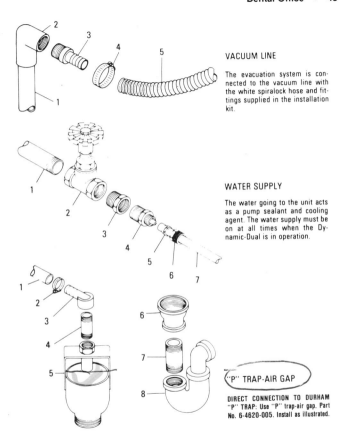

VACUUM LINE

The evacuation system is connected to the vacuum line with the white spiralock hose and fittings supplied in the installation kit.

WATER SUPPLY

The water going to the unit acts as a pump sealant and cooling agent. The water supply must be on at all times when the Dynamic-Dual is in operation.

"P" TRAP-AIR GAP

DIRECT CONNECTION TO DURHAM "P" TRAP: Use "P" trap-air gap. Part No. 6-4620-005. Install as illustrated.

Fig. 4-11. Vacuum system connections.

5
Working Drawings

Fresh-Air Systems

These systems are installed in places where food is sold, stored, handled, manufactured, or processed, such as restaurants, cafes, lunch stands, dairies, bakeries, etc.

Note: Fresh-air master trap shall not be less than 4″ (102 mm) (Fig. 5-1).

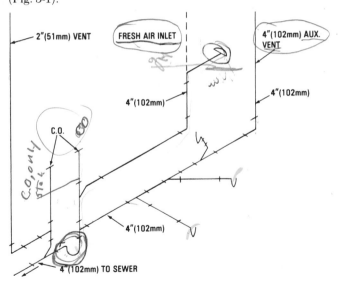

Fig. 5-1. Typical fresh-air system.

Note: Fresh-air inlet may extend through building wall approximately 12″ (31 cm) above grade or extend through roof. This inlet and auxiliary vent shall not be connected to any sanitary vent stack. Check local code.

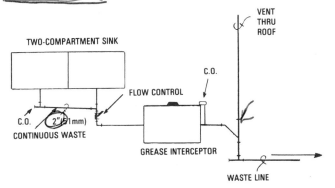

Fig. 5-2. Grease trap installation.

Fig. 5-2 shows a typical grease trap installation. No sink trap is needed when the sink is connected to a grease trap. The interceptor is a trap itself and will prevent sewer gases from entering the house or building.

If a trap was added to this installation, it would restrict the flow and eventually cause a stoppage; not only is this poor practice (double trapping, as it is called), but it is restricted in most plumbing codes.

A vent is provided on the outlet, or sewer side, to prevent siphonage of the contents of the grease trap.

Grease traps should be installed so as to provide access to the cover and a means for servicing and maintaining the trap.

Check the local code in your area, and the procedure set up by the administrative authority.

Electric Cellar Drains

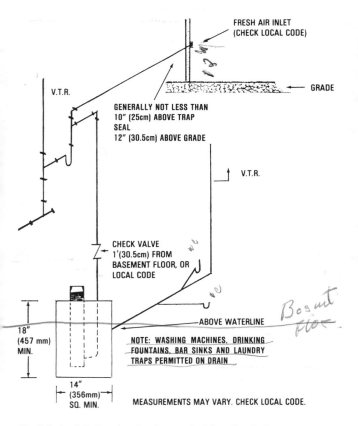

FRESH AIR INLET
(CHECK LOCAL CODE)

GRADE

V.T.R.

GENERALLY NOT LESS THAN
10" (25cm) ABOVE TRAP
SEAL
12" (30.5cm) ABOVE GRADE

V.T.R.

CHECK VALVE
1'(30.5cm) FROM
BASEMENT FLOOR, OR
LOCAL CODE

ABOVE WATERLINE

18"
(457 mm)
MIN.

NOTE: WASHING MACHINES, DRINKING
FOUNTAINS, BAR SINKS AND LAUNDRY
TRAPS PERMITTED ON DRAIN

14"
(356mm)
SQ. MIN.

MEASUREMENTS MAY VARY. CHECK LOCAL CODE.

Fig. 5-3. Installation drawing for an electric cellar drain.

Sand Traps

In the detail in Fig. 5-4, 33½″ (85 cm) would be the distance from the bottom of the pit to the bottom of the inlet or invert.

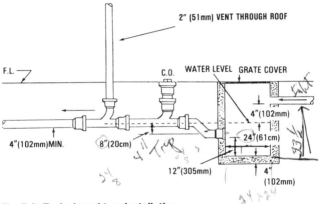

Fig. 5-4. Typical sand trap installation.

Sand traps are generally used in filling stations, garages, poultry houses—places where water carries sand, foam, refuse, or other material that would normally clog an ordinary drain.

Where the sand trap is located in an open area such as a wash rack, slab, or similar place, the 2″ (51 mm) vent may be omitted, Check the local code.

The inside dimension shown in Fig. 5-4 is the minimum size in many areas: 24″ × 24″ (61 × 61 cm).

½ S- or P-Traps

The trap seal is measured from the top dip to the crown weir.

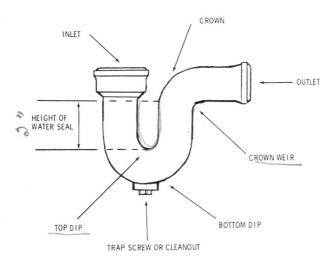

Fig. 5-5. Parts of a P-trap.

To protect a trap water seal from evaporation in a building that will be unoccupied for a period of time, pour a thin film of oil into the trap. During cold seasons, in unheated buildings, water should be drained and replaced with kerosene.

Trap seals may be lost by siphonage, evaporation, capillary attraction, or wind blowing.

The standard P-trap seal is 2" (51 mm); seals over 2½" (64 mm) are called deep seals.

Trailer Connections

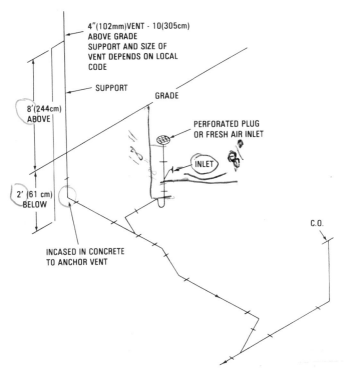

Fig. 5-6. Drawing of rough-in for mobile home or trailer.

The P-trap is at least 18″ (46 cm) below grade, and the inlet not more than 4″ (102 mm) above grade. The connection from the trailer to the inlet should not exceed 8′ (244 cm). The minimum distance between sewer and water connection should be 5' (152 cm). Check the local code.

Acid Diluting Tanks

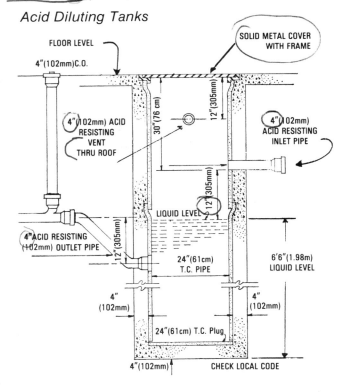

FLOOR LEVEL

SOLID METAL COVER WITH FRAME

4"(102mm)C.O.

4"(102mm) ACID RESISTING VENT THRU ROOF

30"(76 cm)

12"(305mm)

4"(102mm) ACID RESISTING INLET PIPE

12"(305mm)

LIQUID LEVEL

4" ACID RESISTING (102mm) OUTLET PIPE

12"(305mm)

24"(61cm) T.C. PIPE

6'6"(1.98m) LIQUID LEVEL

4" (102mm)

4" (102mm)

24"(61cm) T.C. Plug

4"(102mm)

CHECK LOCAL CODE

Fig. 5-7. Cross-sectional drawing of a typical example of an acid diluting tank installation.

Fig. 5-8 shows three methods of roughing in the waste for a two-compartment sink with a disposal. The first illustration is common in alteration work; it illustrates a double drainage wye inserted into the waste line, so another waste-arm and separate trap to provide for the waste of the disposal can be run.

In the second illustration, the lower sanitary tee is roughed in at 16" (41 cm) to 18" (46 cm) above the finished floor.

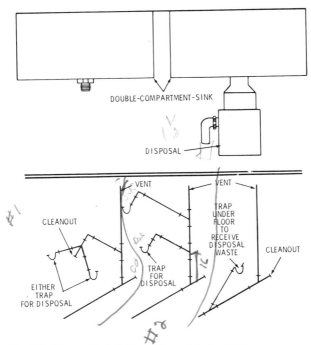

DOUBLE-COMPARTMENT-SINK

DISPOSAL

VENT

VENT

TRAP
UNDER
FLOOR
TO
RECEIVE
DISPOSAL
WASTE

CLEANOUT

CLEANOUT

TRAP
FOR
DISPOSAL

EITHER
TRAP
FOR DISPOSAL

Fig. 5-8. Disposal installation and rough-in drawings.

Check the local code for the proper sizing of the waste pipes, traps, and vent pipes.

Residential Garbage Disposals

Fig. 5-9 shows the parts used to mount a disposal. After the lock ring has been inserted in place, you will then begin

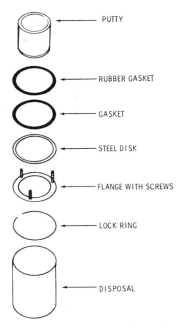

PUTTY

RUBBER GASKET

GASKET

STEEL DISK

FLANGE WITH SCREWS

LOCK RING

DISPOSAL

Fig. 5-9. Exploded view of residential disposal mounting.

tightening the screws in the flange. Do not tighten them all the way until the disposal is set up and positioned for the waste trap.

Once the disposal is mounted and positioned, you can then proceed to secure it to the waste line. If soldering must be done, be sure to tighten all the slip-nuts first and do your sweating or soldering last.

On two-compartment sinks, the disposal is roughed low, around 16" (41 cm), the sink itself around 20" (50 cm) or 21" (53 cm).

No garbage disposal unit shall be installed on any indirect fresh-air waste system or into any grease interceptor.

Cold water must be used with a disposal, as it congeals grease particles, mixing them with food particles being flushed down the drain.

Hot water liquefies grease, and if constantly used, a stoppage would eventually occur due to accumulated coatings of grease.

There is an exception to the above. If a drainage line had just been unstopped by a drain cleaner (see Fig. 2-24, page 16), especially a drain line from a kitchen sink, it is always good practice to flush the newly opened drain with hot water for at least 5 minutes. In this time, all the loosened up grease that had accumulated inside the drain pipe will be liquified by the hot water and immediately flushed down the drain.

6
Roughing and Repair Information

Bathrooms

The following is intended to serve as a general guide for familiarization and for use where rough-in sheets are not available.

Fig. 6-1 shows the general location of fixtures in a bathroom. Typical connections to fixtures are illustrated in Fig. 6-2, and the schematic drawing in Fig. 6-3 shows the sewer and vent layout and connections, using cast iron pipe and fittings with hubs. Of course, local codes should always be checked.

A bathroom that includes one water closet and one 20" (51 cm) lavatory can be placed in a minimum space of 48" (122 cm), finish to finish (Fig. 6-4).

For lavatories (Fig. 6-5), hot and cold water will rough at 20½" (521 mm) if using speedy supplies; if brass nipples are used, check your roughing-in sheets. For the handicapped, the flood rim should be at 36".

The rough-in for waste is 17½" (444 mm) for a pop-up waste or drain plug. The backing ₵ (center line) for the mounting bracket is 31" (787 mm), using a 2" × 10" (5 × 25 cm) support.

The waste location for corner lavatories is 17½" (444 mm) from finished floor, 6¾" (171 mm) from corner to center (left or right). Using speedy supplies, the water location is 20½" (521 mm) from the finished floor, 7" (178 mm) from corner to center; hot on left, cold on right. The ₵ of the backing is 32" (813 mm), using a 2" × 8" (5 × 20 cm) support.

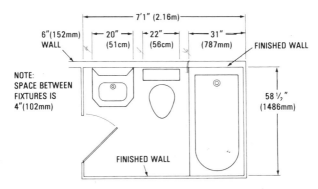

Fig. 6-1. Typical bathroom fixture locations.

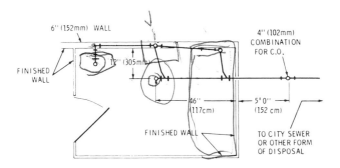

Fig. 6-2. Typical bathroom fixture connections.

For vanitary lavoratories the waste location is not over 16" (406 mm) from finished floor, and the water location not over 18" (457 mm) from finished floor.

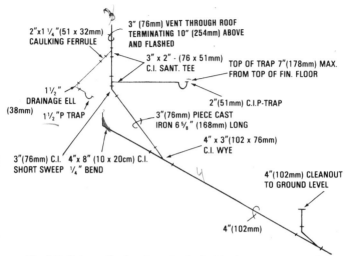

2"x1¼"(51 x 32mm) CAULKING FERRULE

3" (76mm) VENT THROUGH ROOF TERMINATING 10" (254mm) ABOVE AND FLASHED

3" x 2" - (76 x 51mm) C.I. SANT. TEE

TOP OF TRAP 7"(178mm) MAX. FROM TOP OF FIN. FLOOR

1½" DRAINAGE ELL (38mm)

1½ "P TRAP

2"(51mm) C.I.P-TRAP

3"(76mm) PIECE CAST IRON 6⅝" (168mm) LONG

4" x 3"(102 x 76mm) C.I. WYE

3"(76mm) C.I. SHORT SWEEP ¼" BEND

4"x 8" (10 x 20cm) C.I.

4"(102mm) CLEANOUT TO GROUND LEVEL

4"(102mm)

Fig. 6-3. Schematic drawing of a typical bathroom sewer and vent system.

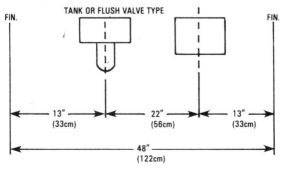

TANK OR FLUSH VALVE TYPE

FIN.

FIN.

13" (33cm)

22" (56cm)

13" (33cm)

48" (122cm)

Fig. 6-4. Lavatory and water closet rough-in dimensions.

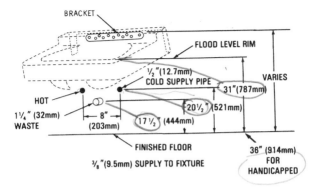

Fig. 6-5. Lavatory rough-in dimensions.

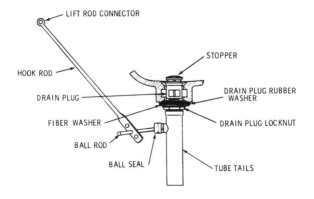

Fig. 6-6. Drawing of a typical pop-up drain.

To install a pop-up drain (Fig. 6-6):

1. Remove the drain plug from the tube tail so as to detach the locknut, rubber washer, and fiber washer.
2. Insert the drain plug through the drain hole of the lavatory, using plumber's putty underneath the flange of the plug. Attach the rubber washer, fiber, or metal washer and locknut.
3. Assemble the tube tail to the drain plug, using a good pipe joint compound, and tighten. Turn the drain so that the side hole in the tube is pointed to the rear of the lavatory. Now, tighten the locknut.
4. Assemble the ball rod assembly to the hole in the side of the tube. Tighten loosely by hand.
5. Insert the stopper into the drain.
6. Attach the hook rod as shown in Fig. 6-6 and tighten the set screw so that the drain works properly by operating the knob.

Kitchen Sinks

The waste line should be 1½" (38 mm) S.P.S., and located 22¼" (565 mm) from the finished floor, 8" (203 mm) off the center line of a double-compartment sink. A single-compartment sink should be roughed in at 25¼" (641 mm). Hot and cold water lines should be 23" (584 mm) from the finished floor: hot 4" (102 mm) to the left of the center line, cold 4" (102 mm) to the right.

If one compartment of a two-compartment sink is to be provided with a garbage disposal, the waste line should rough in at 16" (406 mm) above the finished floor.

Service Sinks

The waste line or trap standard from these sinks is generally 3" (76 mm) S.P.S., and it roughs in at 10½" (267 mm) above the finished floor. The water lines are generally roughed in at 6"

(152 mm) from the finished floor: hot 4" (102 mm) to the left, cold 4" (102 mm) to the right.

Water Closets

Water closet bowls (floor mounted) are generally roughed in at 10" or 12" (254 or 305 mm) from the finished wall on center-line, using a 3" or 4" (76 or 102 mm) S.P.S. waste pipe. Wall-hung water closet bowls will generally rough in at 4½" to 5½" (114 to 140 mm) above the finished floor.

Water closets are supplied with water for flushing by either a flush tank as shown in Fig. 6-7, or a flush valve as shown in Fig. 6-11.

A water closet flush valve supply pipe is located 4¾" (121 mm) to the right from the center line. This cold water supply pipe is 1" (25 mm) S.P.S., 26" (660 mm) from the finished floor, or 20½" (521 mm) from the center of a wall-hung water closet waste line to the center of the water supply line. In hospitals and nursing homes this cold water supply pipe should be roughed in at 36" (914 mm) above the finished floor.

Refer to Fig. 6-7. When flush water is supplied by a flush tank, the water supply piping is usually roughed in 6" (152 mm) off the center line to the left when facing the bowl, and 6" (152 mm) from the finished floor when using speed supplies or soft tubing. The size of the supply pipe should be ½" (13 mm) I.D.

When installing a water closet (floor mounted) to a closet floor flange, whether plastic, brass, or cast iron, be careful as you tighten the closet nuts. Tighten both sides evenly, and tighten just enough so that the bowl does not rock. As soon as the nuts are drawn up snug, sit down on the bowl to settle the wax seal into place, then snug the nuts up a little more. Remember, drawing the nuts up too tightly will crack the bowl!

On the water closet flush valves and lavatory supply tubes, tighten from the top down. This will avoid possible leaks and a waste of time.

To replace a ball cock or float valve in a water closet flush tank (Fig. 6-7):

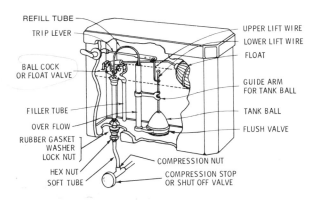

Fig. 6-7. Drawing of a water closet flush tank.

1. Close the valve that supplies water to the tank.
2. Flush the tank and remove the remaining water with a sponge or rag.
3. Holding the float valve with one hand to prevent its turning, begin loosening the hex nut or the nut securing the supply tube to the float valve.
4. Then begin loosening the lock nut and lift the float valve out of the tank.
5. Before placing a new float valve in the tank, be sure the spot is clean and free of dirt or rust.
6. To reassemble, follow instructions in reverse order; be sure to use pipe dope on threads.
7. When a new float valve is installed, take the refill tube, screw it into the opening provided, hold the refill tube at one end, then bend the other end until it enters into overflow pipe.

Referring to Fig. 6-7, you may have occasion to replace a worn or broken flush lever handle that flushes the tank. The first thing you would do is close the valve that supplies the water to the tank. Next, unhook the wire or chain from the old lever. Then remove the flush handle nut inside of the tank (new and old nut are threaded left or counter-clockwise). You can now remove the old handle and lever. Put the new lever in the opening, slide the nut up the lever, large round side first, and bolt the handle in place with the lever in the horizontal position. Last, attach the tank ball wire; in many modern tanks, this is the chain to the Corky, which replaces the tank ball and the lift wires.

Bathtubs and Shower Stalls

To rough in a bathtub (Fig. 6-8), the waste location is 1½″ (38 mm) off the rough wall on the center line of the waste. The P-trap top should be no higher than 7″ (18 cm) below floor level (Fig. 6-8). *Note:* Put the drain piece on last. The shower rod location is 76″ (193 cm) high and 27″ (69 cm) off the finished wall or against the outside tub rim.

Lead pans for shower stalls installed on new concrete floors should be given a heavy coating of asphaltum, both inside and out. The asphaltum protects the lead from corrosion during the curing period of the concrete, due to chemical reaction created between concrete, lead, and the water seeping through and contacting both.

The installation of a typical tub trip waste and overflow is handled as follows. The numbers in parentheses refer to Fig. 6-9.

1. Remove the flat head screw (14) and perforated strainer plate (13) from the drain spud (12). Apply a small amount of putty to the underside of the drain spud.
2. Insert the drain spud (12) through the tub drain outlet from the inside; place the drain spud gasket (11) on the

face of the drain elbow as shown. Proceed to tighten the drain spud (12) until it is secure. When the spud (12) is secure, the drain tube should point directly to the end of the tub. Replace the perforated strainer plate and flat head screw into the drain spud.

3. Place the slip-joint nut and one slip-joint washer on the drain tube or shoe (10), and one slip-joint nut and one slip-joint washer on the riser tube (6). Place the riser tube into the long end of the waste tee (15) and hand-tighten the slip-joint nut. Place the washer (5) on the face of the overflow elbow and push the complete assembly onto the drain tube and hand-tighten the slip-joint nut.

4. Line up the washer with the overflow opening in the tub. Insert the plunger and wire inside the tub and feed the plunger and wire through the opening until the handle and plate line up with the overflow opening in the tub. Secure the plate in place by screwing two oval head screws through the plate into the overflow elbow.

5. Wrench-tighten the two slip-joint nuts so that the drain tube and the riser tube are sealed to the waste tee.

6. The tub may now be placed in position with the tail tube or tailpiece (16) slipped into the drainage line or connected to it and sealed tight.

Depending on the size of the tub, occasionally the drain tube (10) and overflow tube (8) will need to be cut shorter.

Note: Wire and plunger assembly may need to be adjusted so that drain will work properly.

Typical Stall Urinals

The opening left for a urinal should be approximately 24" (61 cm) wide and 18½" (47 cm) from an abutting finished wall.

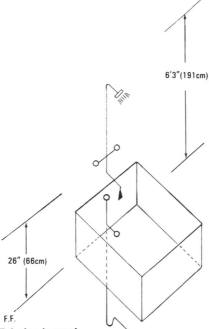

6'3"(191cm)

26" (66cm)

F.F.

Fig. 6-8. Tub view in rough.

The 2" (51 mm) waste line should be ¼" (7 mm) to ⅜" (10 mm) below the top line of the spud or strainer.

The top of the lip should be ¼" (7 mm) below the finished floor. Some codes call for the top of the lip to be above the finished floor (Fig. 6-10). Check the local code.

Sharp sand should be packed under the urinal, and when the urinal is set, sand should be packed at least 1" (25 mm) up on the base of the urinal. When urinals are placed in a battery, spreaders are available; 3" (76 mm) spreaders are popular.

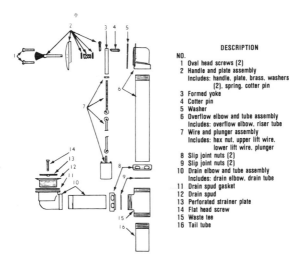

DESCRIPTION

NO.
1 Oval head screws (2)
2 Handle and plate assembly
   Includes: handle, plate, brass, washers
   (2), spring, cotter pin
3 Formed yoke
4 Cotter pin
5 Washer
6 Overflow elbow and tube assembly
   Includes: overflow elbow, riser tube
7 Wire and plunger assembly
   Includes: hex nut, upper lift wire,
   lower lift wire, plunger
8 Slip joint nuts (2)
9 Slip joint nuts (2)
10 Drain elbow and tube assembly
   Includes: drain elbow, drain tube
11 Drain spud gasket
12 Drain spud
13 Perforated strainer plate
14 Flat head screw
15 Waste tee
16 Tail tube

Fig. 6-9. Exploded view of a typical tub trip waste overflow.

Sloan Flush Valves

All Sloan flush valves manufactured since 1906 can be repaired. It is recommended that when service is required, all inside parts be replaced so as to restore the flush valve to like-new condition. To do so, order parts in kit form.

Figs. 6-11 through 6-16 show several views of the Sloan valve.

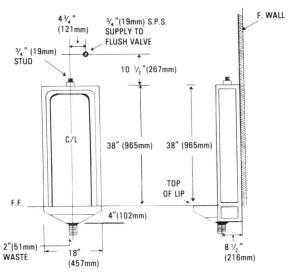

4¾"
(121mm)

¾"(19mm) S.P.S.
SUPPLY TO
FLUSH VALVE

¾" (19mm)
STUD

F. WALL

10 ½"(267mm)

C/L

38" (965mm)

38" (965mm)

TOP
OF LIP

F.F.

4"(102mm)

2"(51mm)
WASTE

18"
(457mm)

8 ½"
(216mm)

Fig. 6-10. Front and side views of a typical stall urinal.

Repairing Water Faucets and Valves

Refer to Fig. 6-17 for parts description.

1. Shut off the water supply valves or stops.
2. Remove the handle screw.
3. Remove the handle (2); be sure the faucet is *not* completely closed before attempting to loosen the locknut (3), or on many faucets a packing nut, that will allow you to remove the stem (5). Open the faucet ¼ turn and continue to check as the locknut is loosened. A crescent wrench should be used.
4. Stem (5) can then be removed.

5. Replace the washer at the bottom of the stem.

   Note: If the Bibb screw holding the washer appears old and difficult to remove, cut the washer out with a penknife. A pair of pliers can then be used, whereas before, a screwdriver may have turned off part of a screw head.

6. Before replacing the stem, examine the seat (10) located at the bottom (inside the valve body) where the washer seats. If the seat (10) appears rough, or a notch or groove is discovered, the seat should be replaced. This seat can be removed by using an Allen wrench in most cases.

    If the seat is not too badly worn, and if the seat is the unremovable type, it can be refaced by using a *seat dressing tool*. The cause of badly worn seats in most cases is delay in replacing worn washers; the passing of water when the valve is shut between washer and seat causes a notch or groove to be worn.

7. When the stem is again inserted into the faucet body, remember to make sure it is kept open slightly. This will prevent damage to the stem.

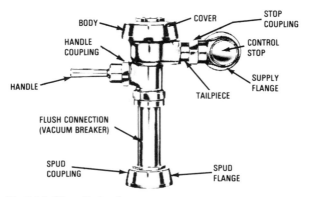

Fig. 6-11. Sloan flush valve.

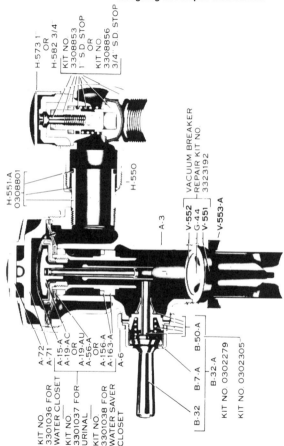

SCREW DRIVER STOP

H-573 1"
OR
H-582 3/4"

KIT NO
3308853
1" S D STOP
OR
KIT NO
3308856
3/4" S D STOP

VACUUM BREAKER
REPAIR KIT NO
3323192

V-552
G-44
V-551
V-553-A

H-551-A
0308801

H-550

A-3

A-72
A-71
A-15-A
A-19-AC
OR
A-19-AU
A-56-A
OR
A-156-A
A-163-A
A-6

B-50-A

B-32-A
B-7-A
B-32

KIT NO
3301036 FOR
WATER CLOSET

KIT NO
3301037 FOR
URINAL

KIT NO
3301038 FOR
WATER SAVER
CLOSET

KIT NO 0302279

KIT NO 0302305

Fig. 6-12. Sloan (diaphragm-type) flush valve.

The repair kits and parts listed are designed to service all Sloan [diagram type] exposed and concealed flush valves. Each item has been identified by a part number along with a corresponding code number. To expedite your replacement requirements order by Code Number.

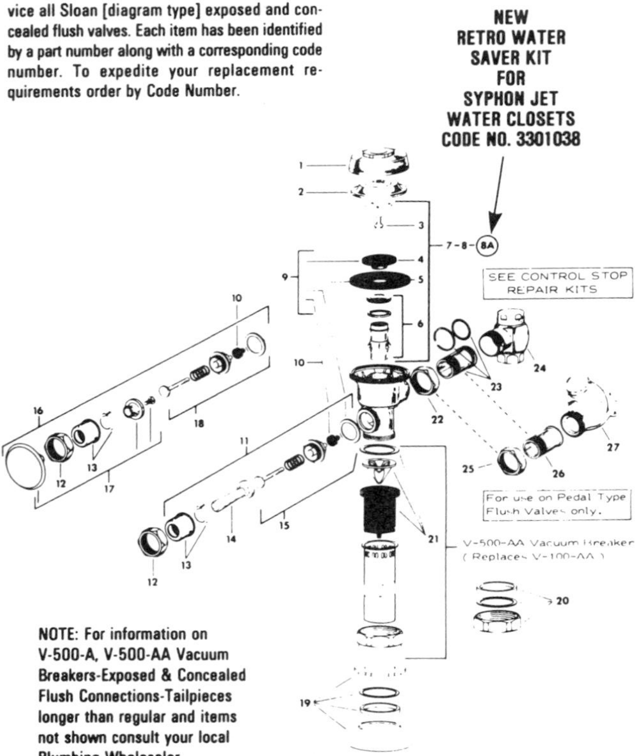

**NEW
RETRO WATER
SAVER KIT
FOR
SYPHON JET
WATER CLOSETS
CODE NO. 3301038**

SEE CONTROL STOP
REPAIR KITS

For use on Pedal Type
Flush Valves only.

V-500-AA Vacuum Breaker
(Replaces V-100-AA)

NOTE: For information on
V-500-A, V-500-AA Vacuum
Breakers-Exposed & Concealed
Flush Connections-Tailpieces
longer than regular and items
not shown consult your local
Plumbing Wholesaler.

Fig. 6-13. Sloan (diaphragm-type) flush valve produced since mid-year 1971.

1.	0301172	*A-72 CP Cover
2.	0301168	A-71 Inside Cover
3.	3301058	A-19-AC Relief Valve (Closet) - 12 per pkg.
	3301059	A-19-AU Relief Valve (Urinal) - 12 per pkg.
4.	3301111	A-15-A Disc - 12 per pkg.
	0301112	A-15-A Disc (Hot Water)
5.	3301188	A-156-A Diaphragm w/A-29 - 12 per pkg.
	0301190	A-156-A Diaphragm (Hot Water)
6.	3301236	A-163-A Guide - 12 per pkg.
7.	3301036	Inside Parts Kit for Closets, Service Sinks, Blowout and Siphon Jet Urinals
8.	3301037	Inside Parts Kit for Washdown Urinals
8A.	3301038	Retro Water Saver Kit-delivers 3- $\frac{1}{2}$ gal.
9.	3301189	A-156-AA Closet/Urinal Washer Set - 6 per pkg.
10.	3302297	B-39 Seal - 12 per pkg.
11.	3302279	B-32-A CP Handle Assem.-6 per pkg.
12.	0301082	*A-6 CP Handle Coupling
13.	0302109	B-7-A CP Socket
14.	0302274	B-32 CP Grip - 12 per
15.	3302305	B-50-A Handle Repair Kit - 6 per pkg.
16.	0303351	C-42-A 3″ CP Push Button Assem.
17.	3303347	3″ CP Push Button Replacement Kit
18.	3303396	C-64-A 3″ Push Button Repair Kit
19.	0306125	F-5-A $\frac{3}{4}$″ CP Spud Coupling Assem.
	0306132	F-5-A 1″ CP Spud Coupling Assem.
	0306140	F-5-A 1-$\frac{1}{4}$″ CP Spud Coupling Assem.
	0306146	F-5-A 1-$\frac{1}{2}$″ CP Spud Coupling Assem.
20.	0306052	*F-2-A $\frac{3}{4}$″ CP Outlet Coupling Assem.
	0306077	F-2-A 1″ CP Outlet Coupling Assem.
	0306092	*F-2-A 1-$\frac{1}{2}$″CP Outlet Coupling Assem. w/S-30
	0306060	*F-2-A 1-$\frac{1}{4}$″CP Outlet Coupling Assem.
	0306093	*F-2-A 1-$\frac{1}{2}$″CP Outlet Coupling Assem.
21.	3323192	V-500-A & V-500-AA Vacuum Breaker Repair Kit
22.	0308676	*H-550 CP Stop Coupling
23.	0308801	*H-551-A CP Adj. Tail 2-$\frac{1}{16}$″ Long
24.	0308757	H-600-A 1″ SD Bak-Chek CP Control Stop
	0308724	H-600-A $\frac{3}{4}$″ SD Bak-Chek CP Control Stop
	0308881	*H-600-A 1″ WH Bak-Chek CP Control Stop
25.	0308889	*H-600-A $\frac{3}{4}$″ WH Bak-Chek CP Control Stop
26.	0308063	*H-6 CP Stop Coupling
27.	0308884	H-650-AG 1″ SD Bak-Chek CP Control Stop
	0306882	*H-650-AG 1″ WH Bak-Chek CP Control Stop

* Items also available in Rough Brass - Consult Local Plumbing Wholesaler for Code Number.

Fig. 6-14. Parts for the flush valve in Fig. 6-13.

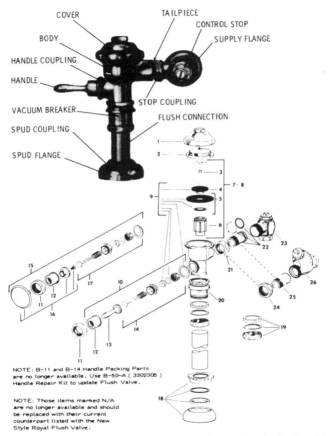

NOTE: B-11 and B-14 Handle Packing Parts
are no longer available. Use B-50-A (3302305)
Handle Repair Kit to update Flush Valve.

NOTE: Those items marked N/A
are no longer available and should
be replaced with their current
counterpart listed with the New
Style Royal Flush Valve.

Fig. 6-15. Sloan (diaphragm-type) flush valve produced prior to mid-year 1971.

1. Cover CP N/A use 0301172 and 0301168
2. Inside Brass Cover N/A use 0301168 and 0301172
3. A-19-A Brass Relief Valve N/A use 3301058 or 3301059
4. 3301111 A-15-A Disc — 12 per package
   0301112 A-15-A Disc (Hot Water)
5. 3301170 A-56-A Diaphragm w/A-29 — 12 per package.
6. Brass Guide N/A use 3301236 — NOTE: 3301236 A-163-A Guide
   replaces all previous Guides.
7. Inside Parts N/A see item No. 7 listed with new style valve — Repair Kit
   replaces all previous inside parts.
8. Inside Parts N/A see item No. 8 listed with new style valve — Repair Kit
   replaces all previous inside parts.
9. 3301176 A-56-AA Washer Set — 6 per package.
10. B-32-A CP Handle Assem. N/A use 3302279
11. A-6 CP Handle Coupling N/A use 0301082
12. B-7 CP Socket N/A use 0302109
13. B-32 CP Grip N/A use 3302274
14. Handle Repair Kit N/A use 3302305
15. C-42-A 3″ CP Push Button Assem. N/A use 0303351
16. 3303347 3″ CP Push Button Replacement Kit
17. 3″ Push Button Repair Kit N/A — use 3303396
18. Spud Coupling Assem. CP N/A — see item no. 19 listed with new style
    valve.
19. Outlet Coupling Assem. CP N/A — see item no. 20 listed with new style
    valve.
20. V-100-A & V-100-AA Vacuum Breaker N/A consult local Plumbing
    Wholesaler for proper V-500-A or V-500-AA Vacuum Breaker
    replacement.
21. H-550 CP Stop Coupling N/A — use 0308676
22. * 0308801 H-551-A CP Adj. Tail 2¹⁄₁₆″ long
23. H-540-A Series Control Stops N/A — see Control Stop Repair Kits or
    item no. 24 listed with new style valve for complete replacement.
24. H-6 CP Stop Coupling N/A use 0308063
25. * 0308026 H-5 CP Ground Joint Tail 1¼″ long
26. H-545-AG Series Control Stops N/A — see Control Stop Repair Kits or
    item no. 27 listed with new style valve for complete replacement.

Fig. 6-16. Parts for the flush valve in Fig. 6-15.

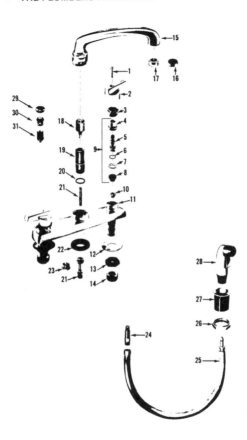

Fig. 6-17. Top-mount Aquaseal® sink fitting with Hermitage® trim and a parts list.

1. Handle Screw
2. Handle
3. Locknut
4. Stem Nut
5. Stem w/Swivel
6. Friction Ring
7. Stop Ring
8. Aquaseal Diaphragm
9. Aquaseal Trim
10. Seat
11. Body
12. Friction Washer
13. Locknut
14. Coupling Nut
15. Spout
16. End Trim
17. Aerator
18. Divertor
19. Post
20. "O" Ring
21. Hose Connection Tube
22. Gasket
23. Body Plug
24. Hose Connector
25. Hose S/A
26. Locknut
27. Spray Holder
28. Spray Head
29. Cap w/Washer
30. Auto Spray (Upper)
31. Auto Spray (Lower)

Note: If part #18 is not used, order parts 29, 30 & 31.

If water leaks out of the handle, it is caused by a worn O ring, a thin rubber ring located on the stem. On some faucets, and on valves such as the globe valve illustrated in Fig. 6-18, the packing nut is the cause of leakage.

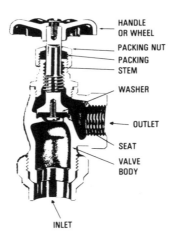

HANDLE OR WHEEL

PACKING NUT

PACKING

STEM

WASHER

OUTLET

SEAT

VALVE BODY

INLET

Fig. 6-18. Drawing of a globe valve.

Open the valve (Fig. 6-15) ¼ turn and tighten the nut snugly; if the valve continues to leak, a new packing washer must be installed. You may also wrap stranded graphite packing around the spindle and tighten snugly. If the spout (15) leaks where it enters body of valve, this is also caused by a worn-out O ring.

The American Standard Aquaseal Valve (Fig. 6-19) has a diaphragm (as illustrated in the Aquaseal kit, shown in Fig. 6-20) in place of a washer.

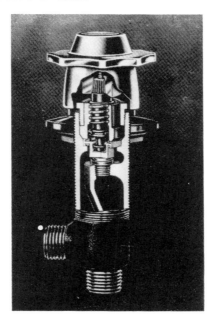

Fig. 6-19. No-drip Aquaseal valve.

Faucets

The secret behind the no-drip feature lies in the washerless valve. All moving parts are outside the flow area, and lubrication on the stem threads is effective for the life of the fitting.

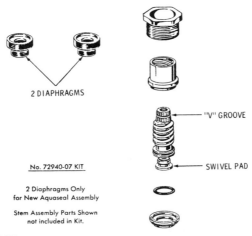

2 DIAPHRAGMS

"V" GROOVE

No. 72940-07 KIT

SWIVEL PAD

2 Diaphragms Only
for New Aquaseal Assembly

Stem Assembly Parts Shown
not included in Kit.

Fig. 6-20.

There is no seat washer wear, which is the cause of leaks and
dripping in ordinary fittings.

When it becomes necessary to replace the diaphragm in an
Aquaseal valve, remove the handle and check for a V groove
around the stem, located in the middle of the splines. If you *do
not see* the V groove, replace the handle and contact your
supplier for the appropriate stem assembly. If the V groove is
visible, proceed to remove the valve unit.

After removing the old diaphragm, turn the stem so that one
thread still protrudes from the top of the stem nut. Slip a new
diaphragm over the swivel pad and insert assembly into the
fitting, exercising normal care to prevent damaging the diaphragm. Tighten the valve unit and replace the handle.

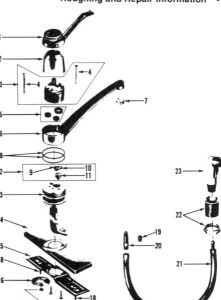

Fig. 6-21. Single-control Aquarian® sink fitting.

1. Hands
2. Escutcheon Cap
3. Cartridge (4 Gal.)
4. Cartridge Screw
5. Cartridge Seal Set
6. Spout S/A w/Aerator
7. Aerator
8. O Ring
9. Divertor
10. O Ring
11. Retainer S/A
12. Diverter & Retainer Set

13. Manifold (4 Gal.)
14. Escutcheon
15. Mounting Plate
16. Washer Slotted
17. Nut
18. Carriage Bolt
19. Pipe Plug
20. Hose Connection
21. Hose S/A
22. Locknut & Spray Holder
23. Spray Head S/A

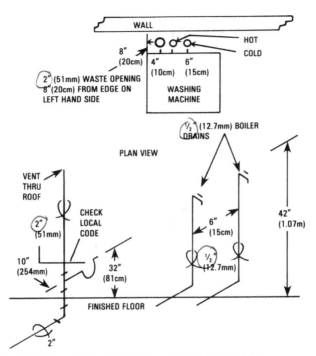

Fig. 6-22. Rough-in specifications for a washing machine.

Fig. 6-21 is an exploded view of the American Standard single-control Aquarian sink fitting.

Domestic Washing Machines

The drawings in Fig. 6-22 show rough-in specifications and a schematic diagram of water and waste lines running to and from a washer. Be sure, of course, to check local codes.

7
Outside Sewage Lift Station

Outside Sewage Lift Station Piping Illustrated

The following illustrations and related data are intended to serve as a guide, and to familiarize the plumber or fitter with the most important features of a typical sewage lift station installation.

Fig. 7-1 features a cutaway drawing of an outside sewage lift station. Fig. 7-2 shows the discharge piping extending out through the top of the ejector. Fig. 7-3 illustrates the piping arrangement when the discharge piping comes from the side of the basin. Fig. 7-4 shows a cross-sectional view of the pump with reference numbers for construction features. Fig. 7-5 is a cross-sectional reference drawing for guide bearing, etc. Figure 7-6 is a reference drawing for adjusting the impeller in the pump featured herein, if it ever became necessary. Fig. 7-7 features the mercury float switch, which is used in conjunction with duplex pump applications. Fig. 7-8 features the fiberglass outside pump housing used in this illustration. It is a lightweight fiberglass, recommended for outdoor installations, and is constructed of high strength, polyester resin reinforced with glass fibers. Durable weatherproof construction is assured, and painting is not necessary. Because of the housing's light weight, it is very easy to install or remove when it becomes necessary for a major pump overhaul.

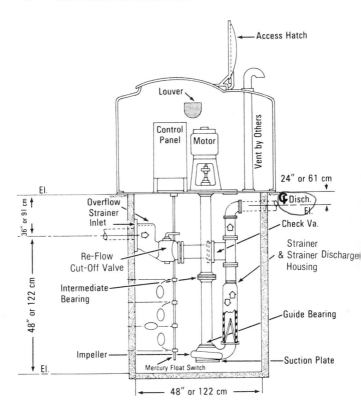

Fig. 7-1. Sewage lift station.

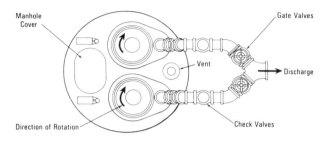

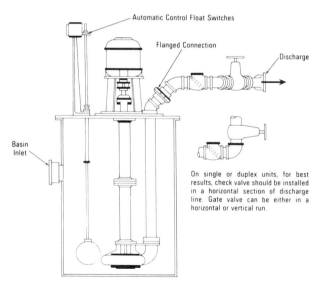

On single or duplex units, for best results, check valve should be installed in a horizontal section of discharge line. Gate valve can be either in a horizontal or vertical run.

Fig. 7-2. Discharge piping layout.

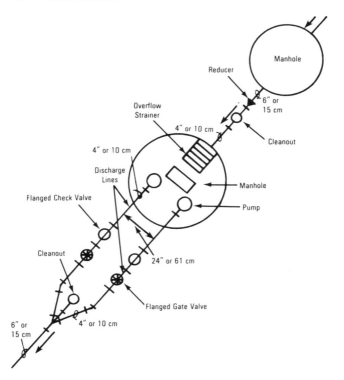

Fig. 7-3. Piping discharging from side of basin.

General Instructions for Vertical Pumps

Vertical centrifugal pumping units, if installed, are very simple to maintain.

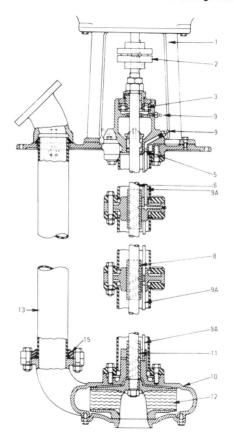

Fig. 7-4. Cross-sectional view of pump.

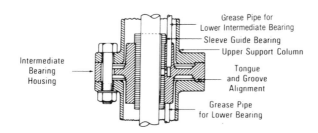

Fig. 7-5. Cross-sectional reference diagram.

Before installing these pumps in a basin, make sure that the basin is fairly clean.

Any accumulation of sand, dirt, cinders, etc., should be cleaned out, or unnecessary wear to the pump will occur. If solid matter is allow to accumulate, it will gradually close off the suction of the pump. It is far more expensive to replace worn parts than to clean the basin at regular intervals.

Refer to Fig. 7-6. The impeller in this pump must be held in the center of the space provided for it in the pump casing and must not rub against the casing. Turn the pump shaft by hand. If the shaft does not turn freely, that is an indication that the impeller is rubbing. A micrometer adjustment is provided at the ball thrust bearing to raise or lower the shaft and impeller to the proper position. Do not change this adjustment unless it is necessary. To make an adjustment, loosen the set screw in the ball bearing thrust collar and back off the adjusting nut slightly. This nut is a combination adjusting and locknut. It fits tight around the shaft threads and will offer some resistance in turning. When backing off the nut, turn the shaft, pressing downward so the impeller rubs on the suction plate. Take up the nut until the impeller just clears the suction plate and the shaft turns freely. Retighten the set screw in the thrust collar.

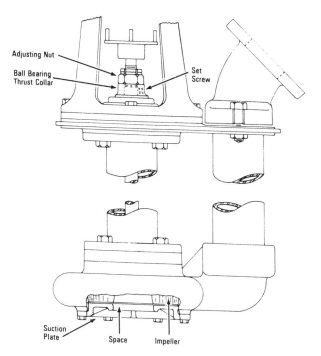

Fig. 7-6. Reference diagram for adjusting impeller.

Before turning on the current, be sure the shaft rotates freely. Check the direction of rotation — see the arrow on the pump floor plate.

For ball bearings and guide bearings, it is well to add two or three ounces (57 or 85 grams) of grease at a time, at regular

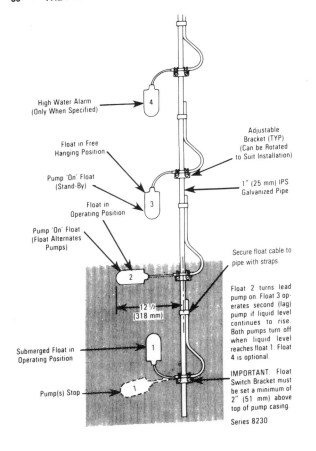

High Water Alarm
(Only When Specified)

Float in Free
Hanging Position

Pump 'On' Float
(Stand-By)

Float in
Operating Position

Pump 'On' Float
(Float Alternates
Pumps)

Submerged Float in
Operating Position

Pump(s) Stop

Adjustable
Bracket (TYP)
(Can be Rotated
to Suit Installation)

1" (25 mm) IPS
Galvanized Pipe

Secure float cable to
pipe with straps.

Float 2 turns lead
pump on. Float 3 op-
erates second (lag)
pump if liquid level
continues to rise.
Both pumps turn off
when liquid level
reaches float 1. Float
4 is optional.

IMPORTANT: Float
Switch Bracket must
be set a minimum of
2" (51 mm) above
top of pump casing.

Series 8230

12½
(318 mm)

Fig. 7-7. Mercury float switch.

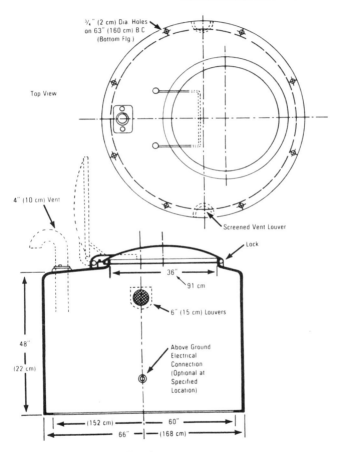

Top View

¾" (2 cm) Dia Holes
on 63" (160 cm) B C
(Bottom Flg)

Screened Vent Louver

4" (10 cm) Vent

Lock

36"

91 cm

6" (15 cm) Louvers

Above Ground
Electrical
Connection
(Optional at
Specified
Location)

48"

(22 cm)

(152 cm)   60"
66"   (168 cm)

Fig. 7-8. Outside pump housing.

intervals, until it can be determined how often more grease will be needed. The grease tubes for lubricating the lower guide bearings are enclosed in the hanger pipes.

Pump Construction Features (Fig. 7-4)

All joints in the pump suspension system are tongue and groove type. The intermediate bearings and bearing housings are self-contained assemblies that ensure ease of maintenance and proper fit and alignment after dismantling.

1. *Motor support.* Top end machined to match the NEMA C motor end flange. No shims are necessary for perfect alignment.
2. *Flexible coupling.* Properly sized and designed for the pump load and the motor speed.
3. *Ball thrust bearing.* Located in a sealed housing and protected from dirt. It is of ample capacity to carry the weight of the pump shaft and impeller.
5. *Stuffing box* with packing.
6. The *pump shaft* is of ample diameter in all pump sizes to prevent any whipping action. It provides a large factor of safety for handling maximum loads and shocks in pumping unscreened sewage.
8. *Guide bearing* (Fig. 7-5).
9. *Pressure grease fittings.* For all bearings.
9A. *Lube pipe.* Located in the shaft support column, except where a large number of intermediate bearings in a pump over 12′ (3.66 m) long makes it necessary to locate it outside.
10. *Pump casing* is cast iron joined to the steel suspension pipe with tongue and groove flange for rigid, permanent alignment. Note the wide, smooth passages permitting free flow at all points.
11. *Casing bearing* located close to the impeller hub to reduce overhang. It is designed for maximum wearing

surface and equipped with a forced feed grease line for oil lubrication as specified. Forced grease is recommended.

12. *Impeller.* The most important single part of the sewage pump. Here is where experience counts, and it is important to obtain pumps with the best quality impellers. The impeller is tight-fitting, held in rotation with a stainless steel shaft key, and dynamically balanced.

13. *Discharge pipe.*

15. *Expansion fitting.* Prevents distortion and strain on the shaft and bearings.

8

Pipes and Pipelines

Pipelines Hung from a Ceiling

Instead of running pipe in a ceiling, where possible, sleeve locations should be transferred from the floor above to the floor below. Make use of columns and walls to square off your work. Once points to be reached are established and you know in which direction you are heading and where you are coming from, it will become clear what fittings will work. Cuts can then be determined, and even the location of hangers, if inserts were not provided in the building construction.

The next step is transferring these hanger locations to the ceiling by way of your plumb bob.

When running hot and cold water headers to a number of fixtures in the same area, especially a battery of lavatories, the headers should be run as shown in Figs. 8-1, 8-2, and 8-3. Water supply pipes extending vertically one or more stories are called risers. Soil and vent pipes extending vertically are called stacks. A 2% grade is slightly less than ¼″ per foot.

Examples:

A sewer 220 feet long:

$$
\begin{array}{r}
220 \\
\times\ .02 \\
\hline
\end{array}
$$

4.40′ or approximately 52¾″ — total fall

In metric — a sewer 37 meters long:

$$
\begin{array}{r}
37 \\
.02 \\
\hline
\end{array}
$$

.74 meters, or 740 mm — total fall in 37 meters on 2% grade

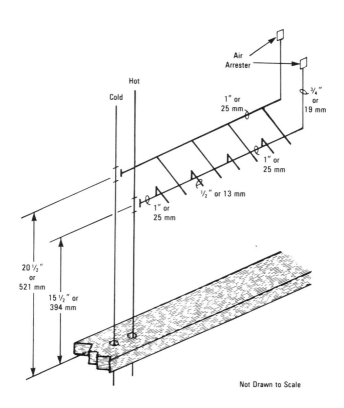

Fig. 8-1. Hot and cold water headers for lavatories.

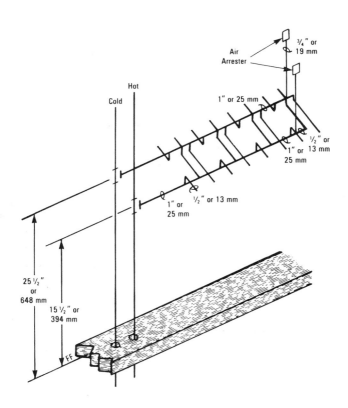

Fig 8-2. Hot and cold water headers for back to back lavatories.

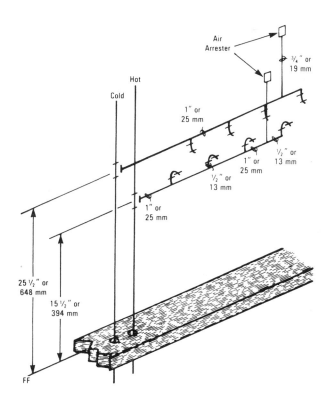

Fig. 8-3. Schematic of hot and cold water headers.

When a line is neither horizontal nor vertical, it is said to be slanting or diagonal.

A 1" (25 mm) water pipe is equal to four ½" (13 mm) pipes.

Eighteen inches (46 cm) of 4" (10 cm) pipe holds one gallon (3.785 liters) of water.

Pipe above 12" (305 mm) diameter is generally classified by its outside diameter. Thus, 14" (356 mm) pipe would be its O.D.

A *cross-connection* is any physical connection or arrangement of piping that provides a connection between a safe water supply system and a separate system or source that is unsafe or of questionable safety and which, under certain conditions, permits a flow of water between the safe and unsafe systems or sources.

Air chambers, to be effective, should be located as close as possible to the points at which *water hammer* will occur. Water hammer becomes much greater at 100 psi (689.5 kPa) than at 50 psi (345 kPa). Two types of manufactured devices used to reduce or eliminate water hammer are *shock absorbers* or *water hammer arrestors* to cushion the shock and *pressure-reducing valves* to lower the operating pressure (Fig. 8-4).

Condensation is formed on cold water pipelines when warm, humid air comes into contact with the cold surfaces, causing these pipelines to give up some of their moisture.

A *corporation stop* is located at the tap in the city water main. The connection between the city water main and the building is called the *service pipe.*

The *invert* of a sewer line or any pipeline is the inside wall flow line at the bottom of the pipe.

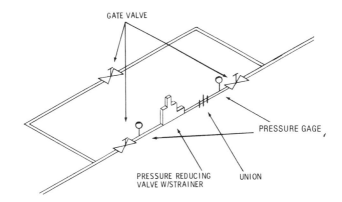

GATE VALVE

PRESSURE GAGE

PRESSURE REDUCING
VALVE W/STRAINER

UNION

Fig. 8-4. Installation of pressure-reducing valve.

Brass pipe expands about 1¼" (32 mm) per 100' (30.5 meters) for a 100°F (38°C) rise in temperature.

Steel pipe expands about ¾" (19 mm) per 100' (30.5 meters) for a 100°F (38°C) rise in temperature.

When pressure exceeds 80 psi (552 kPa), a pressure regulator should be installed, especially where such pressure leads to a hot water heater. Check the local code.

Connections Between Heater and Storage Tank with Bypass

See Fig. 8-5 for details.

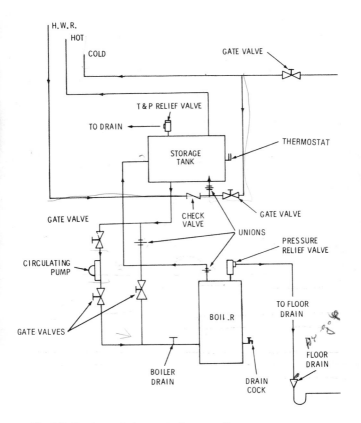

Fig. 8-5. Heater and storage tank connections.

9
Vents, Drain Lines, and Septic Systems

Definitions

A *continuous* vent (also back vent) is a vertical vent that is a continuation of the drain to which it connects. The *main* vent, same as *vent stack*, is the principal artery of the venting system to which vent branches may be connected. A *branch* vent is a vent pipe connecting one or more individual vents with a vent stack or stack vent. A *wet* vent is a waste pipe that also serves as an air circulating pipe or vent.

A *circuit* vent is a branch vent that serves two or more traps and extends from in front of the last fixture connection of a horizontal branch to the vent stack. An *individual* vent is a vent pipe installed to vent a fixture trap. It may connect with another vent pipe 42 inches (1.07 m) or higher above the fixture served or terminate through the roof individually. A *dual* vent, also called *common* or *unit* vent, is a vent connecting at the junction of two fixture drains and acting as a vent for both fixtures. A *relief* or *yoke* vent is a vent where the main function is to provide circulation of air between a drainage and vent system.

A *local* vent is a ventilating pipe on the fixture inlet side of the trap; this vent permits vapor or foul air to be removed from a fixture or room. This removal of foul air or offensive odors from toilet rooms is accomplished now by bathroom ventilation fans and ducts. A *dry* vent conducts air and vapor only to the open air. A *loop* vent is the same as a circuit vent, except that it loops back and connects with a stack vent instead of a vent stack.

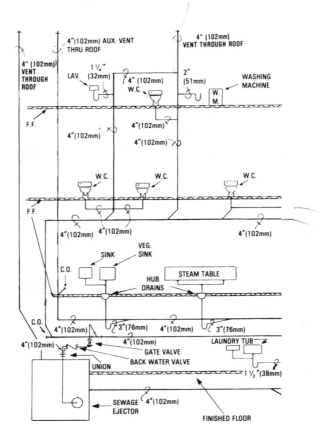

Fig. 9-1. Public building vent and drainage system drawing.

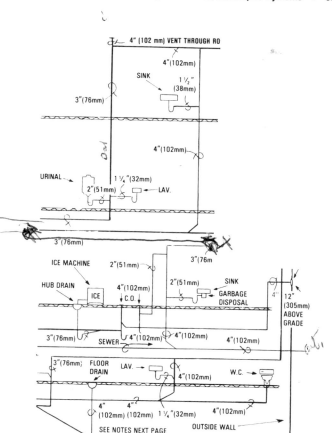

4" (102 mm) VENT THROUGH RO

4"(102mm)

SINK

1 1/2" (38mm)

3"(76mm)

4"(102mm)

URINAL

1 1/4"(32mm)

2"(51mm)

LAV.

3"(76mm)

3"(76m)

ICE MACHINE

2"(51mm)

HUB DRAIN

4"(102mm) C.O.

2"(51mm)

SINK
GARBAGE
DISPOSAL

ICE

4" 12" (305mm) ABOVE GRADE

3"(76mm)

SEWER   4" (102mm)   4"(102mm)   4"(102mm)

3"(76mm)   FLOOR DRAIN

LAV.

4"(102mm)

W.C.

4" (102mm) (102mm)   1 1/4"(32mm)   4"(102mm)

SEE NOTES NEXT PAGE

OUTSIDE WALL

Purpose of Venting in a Drainage System

The purpose of venting is to provide equal pressure in a plumbing system. Venting prevents pressures from building up and causing retarded flow. It protects trap seals and carries off foul air, gases, and vapors that would form corrosive acids harmful to piping.

A plumbing system is designed for not more than 1" (2.5 cm) of pressure at the fixture trap. Greater pressures may disturb the trap seals. One inch (2.5 cm) of pressure is equivalent to that of a 1-inch column of water.

A waste stack terminates at the highest connection from a fixture. From this point to its terminal above the roof, it is known as a stack vent. In colder climates, the closer the vent terminal is to the roof, the less chance of frost closure.

Examples of Venting

Figs. 9-1 through 9-11 illustrate a number of venting situations, including a residence in Fig. 9-2. The system in Fig. 9-1 delivers fresh air for an ice machine, steam table, vegetable sink, and powder rooms. Figs. 9-3, 9-4, 9-5, 9-6, 9-7, and 9-9 show vent systems for water closets, tubs, and lavatories. The drawing in Fig. 9-8 shows combination and vent stacks, and Fig. 9-10 illustrates a "looped vent" with a bleeder. In all cases, walls and partitions have been eliminated for simplicity. Where sewage-ejector and fresh-air vents are used, both must extend through the roof independently. A garbage disposal must empty directly into the sanitary system, not the fresh-air system. The horizontal branch waste line should be 3" (76 mm) and continue up to the sink opening. Check local code in all cases. The following common abbreviations are used in the drawings: C.O. = cleanout; W.C. = water closet; V.T.R. = vent through roof.

If a garbage disposal is installed on a sink, the horizontal branch waste line should be 3" (76 mm) and continue up to the sink opening. Check local code.

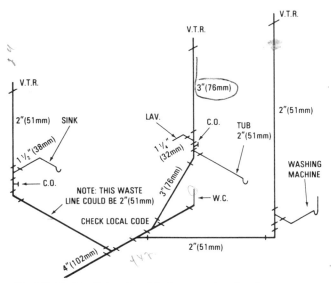

Fig. 9-2. Residential vent and drainage system drawing.

Septic Tanks

A septic tank will work properly for many years if installed correctly. A septic tank should be made water and air tight, so that the bacterial action that disintegrates the solid matter can take place.

Septic tanks are designed to operate full while the bacterial action works on the solids, the liquid overflow drains off into the drain field.

The tank itself does not need a cleanout, since the pre-cast slabs can easily be removed if ever it becomes necessary to clean the tank. Once a tank is cleaned, and pre-cast labs are set

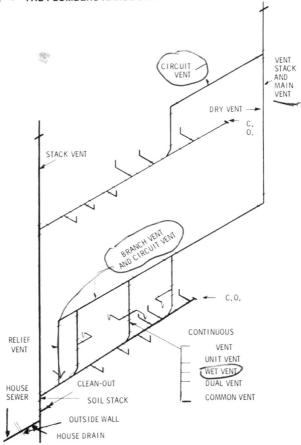

Fig. 9-3. Drawing of water closet and lavatory vents.

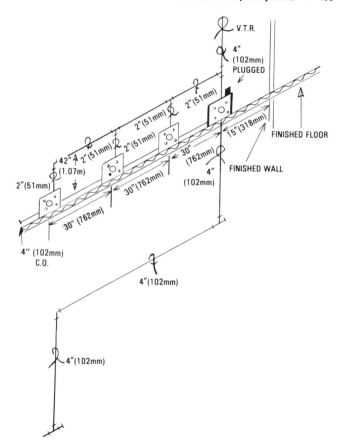

Fig. 9-4. Venting system for wall-hung water closets.

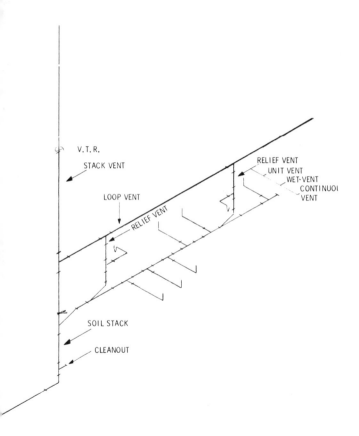

Fig. 9-5. Another water closet and lavatory venting system.

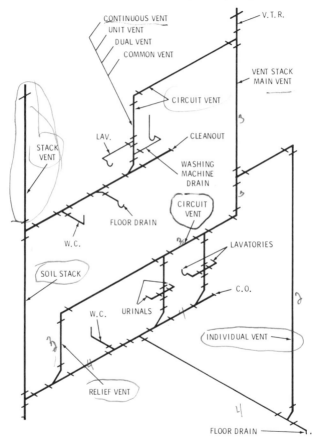

Fig. 9-6. Venting system for wall-hung urinals, floor drains, and water closets.

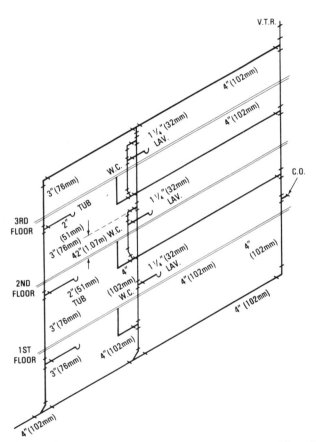

Fig. 9-7. Tubs, water closets, and lavatories are vented in this system.

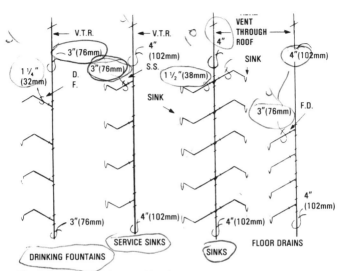

Fig. 9-8. System using combination waste and vent stacks.

back into place, they can then be resealed with cement or mortar-mix minus sand or gravel.

Drain fields can and will become spent in time, and new drain lines must be laid. These new lines must begin at the distribution box and extend out in new directions.

Information may vary according to location and type of soil. Check local code.

The following data is for septic systems in hard compact soil:

2-bedroom house — 750 gallon (2839 liter) capacity with 200 feet (61 m) of drain field.

3-bedroom house — 1000 gallon (3785 liter) capacity with 300 feet (91.5 m) of drain field

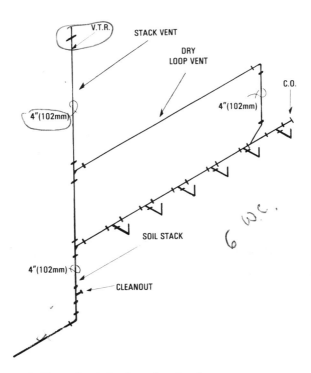

Fig. 9-9. Water closets in a looped vent system.

4-bedroom house — 1000 gallon (3785 liter) capacity with 400 feet (122 m) of drain field

In sandy soil smaller tanks and less drain field footage are generally the rule.

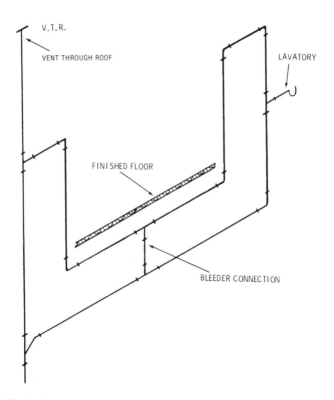

Fig. 9-10. Looped vent using a bleeder.

The top of tank is generally 6 inches (15 cm) to 10 inches (25 cm) below ground level. The distance from the inlet invert (or

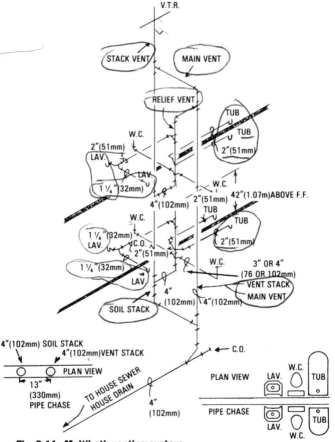

Fig. 9-11. Multibath venting system.

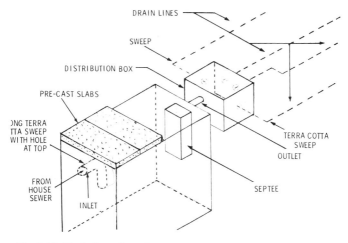

Fig. 9-12. Typical septic tank installation.

bottom part of inlet) to the septic tank from the top of the tank is generally 12 inches (305 mm). The outlet invert is generally 2 inches (51 mm) lower.

If drain lines must run parallel to each other, these lines should be at least 10 feet (3.05 m) apart. Terra cotta or cement drain tile is used, measuring 4 inches (102 mm) inside and 12 inches (31 cm) long.

Wastewater Treatment Plants

The use of safe, live, nontoxic bacteria overcomes wastewater treatment problems in residential, commercial, and municipal wastewater treatment systems.

Approved for waste-related applications by USDA, new BIO JET-7® is used to reduce hydrogen sulfide in gravity systems and to reduce organic line deposits. This process leads to the elimination of the odor problem. Because BIO JET-7 can reestablish and enhance normal biological activity in wastewater treatment systems, this product stops organic odors and prevents them from recurring, as long as the maintenance program is followed.

Other advantages of this type of biological treatment are its ability to prevent clogging in leaching fields and beds; to enhance the biological activity in sewers, drain lines, treatment plants, septic tanks, and disposal fields and beds; to eliminate the threat of sewage backup and to reduce sludge deposits.

There are usually not enough natural bacteria in wastewater treatment systems to decompose the waste properly before it leaves the system. Man-made wastewater treatment systems suffer from high concentrations of organic waste, insufficient bacteria population and the absence of the right kind of bacteria to decompose organic wastes.

The results are noxious odors, sludge buildup, sewage backups, wastes that are discharged before they are completely decomposed, facilities that are overloaded with organic waste, and a lot of expensive facility repair, disposal field or bed replacement and pollution cleanup.

Packaged Wastewater Treatment Plants

Pollution control for subdivisions, motels, factories, schools, restaurants, service stations, mobile home parks, and hundreds of other facilities beyond city sewer lines is possible with the commercial wastewater treatment plant. It is engineered for minimum maintenance and long life (Fig. 9-13).

The packaged wastewater treatment plant solves wastewater problems. It makes it possible for motels and service stations to be built along interstate highways far from towns, subdivisions

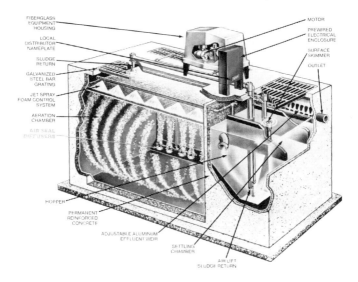

FIBERGLASS
EQUIPMENT
HOUSING

MOTOR

LOCAL
DISTRIBUTOR
NAMEPLATE

PREWIRED
ELECTRICAL
ENCLOSURE

SLUDGE
RETURN

SURFACE
SKIMMER

OUTLET

GALVANIZED
STEEL BAR
GRATING

JET SPRAY
FOAM CONTROL
SYSTEM

AERATION
CHAMBER

AIR SEAL
DIFFUSERS

HOPPER

PERMANENT
REINFORCED
CONCRETE

ADJUSTABLE ALUMINUM
EFFLUENT WEIR

SETTLING
CHAMBER

AIR LIFT
SLUDGE RETURN

Fig. 9-13. Packaged wastewater treatment plant. *(Courtesy of JET, Inc.)*

to be planned in scenic areas miles beyond sewer lines, and factories to be located on outlying sites.

Plumbers should be aware of this type of plant so that they can readily advise those who wish to start planning on upgrading an existing system or installing a new one. Installation plans come with the purchase of the packaged system. However, there are some bits of information the plumber should know about this particular plant.

This is a three-stage treatment process: pre-treatment, aeration, and settling.

Pre-treatment. In this JET plant (Fig. 9-13), large objects in the wastewater are caught by pre-treatment devices such as bar screens, trash traps, or comminutors (wastewater grinders) and broken down before being allowed to pass into the aeration chamber. Untreatable material like plastic or metal is kept out completely.

Aeration. After pre-treatment, the wastewater flows into an aeration tank where it is mixed with air. Air diffusers at the bottom of the aeration tank bubble in large amounts of air for two purposes — to meet the oxygen demand of the aerobic digestion process and to mix the aeration tank contents, insuring complete treatment. In the aeration tank, the pre-treated wastewater is held for 24 hours while being transformed into a clear odorless liquid.

Settling. From the aeration tank the treated liquid flows into a settling tank that holds the liquid completely still. Here any small particles in suspension settle to the bottom and are returned to the aeration tank for further treatment.

This settling process in the final tank of the plant leaves a clear, highly treated water at the top. Only this highly treated liquid (called *effluent*) leaves the plant and returns to the environment.

A 100,000-gallon-per-day packaged wastewater treatment plant is shown in Fig. 9-14. It is located in a subdivision of Little Rock, Arkansas. Note the fiberglass housings for the motor and electrical controls.

Beginning Drain Lines

After the tank is set, you begin the drain field by placing one drain tile between the outlet opening and the distribution box.

At each opening leading to a drain line a ditch approximately 24″ (61 cm) wide is dug 6″ (51 cm below the outlet opening of the distribution box; wooden pegs are then driven in the ground,

---

*This information furnished by: JET, Inc., 750 Alpha Drive, Cleveland, OH 44143

Fig. 9-14. A 100,000-gallon-per-day packaged plant installed to serve a large city's subdivision. *(Courtesy of JET, Inc.)*

beginning at each outlet of the box and spaced every 12″ to 18″ (31 to 46 cm) apart; top of peg should be level with the bottom of the outlet openings in the distribution box. Pegs should be laid level or pitched approximately 1″ (25 mm) in 100′ (30.5 m).

Next, the crushed rock or gravel is installed to the level of the pegs or 6″ (15 cm).

Now begin laying the drain tile, being careful to space each tile approximately ³⁄₈″ (10 mm) apart; these cracks or spaces are to be covered with tar paper.

After the tile is set, each drain line will then receive more crushed rock until rock reaches 1″ (25 mm) above the drain tile.

The last step is to cover the entire drain line with tar paper, then cover with earth.

Note: A 1000-gallon (3785-liter) septic tank contains approximately 133¾ cubic feet (3.787 cubic meters) of space. This size tank could measure

> 42″ (107 cm) wide
> 84″ (214 cm) long
> 66″ (166 cm) deep

A 750-gallon (2839-liter) septic system contains approximately 100¼ cubic feet (2.838 cubic meters) of space. This size tank could measure

> 35″ (89 cm) wide
> 72″ (183 cm) long
> 68¾″ (175 cm) deep

Measurements are inside dimensions.

10
Lead Work

A plumber should buy only good wiping solder that has the manufacturer's name and brand indicating composition cast on the bar. Refusal to accept anything less will eliminate many solder troubles that can and do occur. Good wiping solder usually contains between 37% and 40% pure tin, between 63% and 60% pure lead.

Complete solidification occurs around 360°F (182°C); complete liquefaction around 460°F (238°C). In general, the working range will be about 100°F (38°C).

Many plumbers twist a piece of newspaper and dip it into molten solder to check for correct wiping heat. When the paper scorches but does not ignite, the solder is hot enough.

Preparing Horizontal Round Joints

First, the ends of the pipe to be joined should be squared off with a coarse file or rasp and the pipe should be drifted so it is uniformly round and free from dents.

Grease and dirt should be cleaned off the surface of the pipe for about 4″ (102 mm) from the ends. With a knife, ream out burrs in the ends of the pipe to be joined.

Then one end should be flared, using a turnpin and mallet, until the inside diameter at the end equals the original outside diameter. The end thus flared is the one in which the water will flow. The shoulder or outside edge of the flare should then be rasped off (Fig. 10-1) approximately parallel with the outside wall of the pipe. An adequate flare for lead pipe is ¼″ (6.4 mm) to ⅜″ (9.5 mm).

The inside of the flared end should be soiled for about 1" (25 mm). Next, the end of the other section of pipe, which will be the one from which the water will flow, should be beveled with a rasp until it fits snugly inside the flared end. In this way, the joint is made in direction of the flow, reducing resistance and chance of clogging. With dividers, mark a line around the flared end of the pipe, at a distance from the extreme end, equal to half the length of the finished joint (half being generally $1\frac{1}{4}$" or 32 mm).

Mark a similar line on the beveled end at a distance from the extreme end equal to half the length of the joint, plus the length of the beveling. This will make the center of the finished joint at the intersection of the two outside surfaces of pipe. Next mark an additional 3" (76 mm); in this 3" (76 mm) portion, which will be at the extreme ends of both pieces of pipe to be joined, clean lightly with wire brush, dust off, and apply plumber's soil.

Next, the $1\frac{1}{4}$" (32 mm) portions on each pipe totaling $2\frac{1}{2}$" (64 mm), the flare and bevel should be lightly scraped clean (Fig. 10-2) with a shave hook and immediately covered with a thin coating of tallow to prevent oxidation.

Now the ends of pipe should be fitted snugly together and braced in position (Fig. 10-3) so they will be absolutely stationary during and after wiping until the solder has cooled.

The bottom of the pipe should be about 6" (152 mm) above the bench or working place. When wiping joints in place, if the space is more than 6" (152 mm), a box or some other flat object should be placed so there will be a surface about 6" (152 mm) under the joint to prevent splashing of solder.

To aid in getting heat up on a joint quickly, one or both of the outer, or extreme ends, of the pipe should be plugged; usually newspaper is used. Proper heat is 600°F (316°C).

After the wiping solder is heated, carefully stirred, and skimmed to remove dross, tested for proper heat (as described previously), wiping may begin. Wiping cloths are usually made of 10 oz. (284 gram) herringbone material. The generally

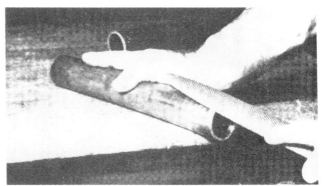

Fig. 10-1. Beveling the male end of lead pipe with a rasp preparatory to fitting it into the flared end for wiping. A close fit is highly important to successful joint wiping.

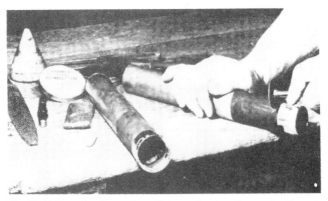

Fig. 10-2. Scraping the male end clean after beveling and soiling. Female end, at left, flared, soiled, and cleaned, is ready to have the male end inserted in it.

Fig. 10-3. Pipes prepared and fitted together ready for wiping. Note how they are held securely by boards and bricks with small boards under one side to prevent rolling. The paper under the joint is to catch excess solder.

Fig. 10-4. Solder has just been poured on the joint and that caught in the cloth is being pressed against the bottom to get up heat.

used wiping cloth sizes are 3″ (76 mm) cloth for a 2½″ (64 mm) joint, 3¼″ (83 mm) cloth for a 2¾″ (70 mm) joint, measured by length of joints. Figs. 10-4 to 10-7 show solder being applied to a joint.

Procedure for Cleaning Wiping Solder

The procedure for cleaning wiping solder varies. One method is: heat solder to a dull red, about 790°F (421°C) (melting point of zinc), then add about one tablespoon of sulphur and stir. Then permit the pot to cool slowly and skim off top dross that contains the impurities consisting of compounds of lead, tin, and zinc. Stir and skim until the top is clean. Next, add a small amount of powdered or lump rosin, stir, and skim again. Allow the pot to cool until the solder reaches wiping temperature; then add sufficient tin to reestablish proper workability.

Wiping Head Stub on Job Brass Ferrule

Prepare by using a fine file, apply Nokorode, and proceed to tin the ferrule.

For the lead stub use a fine file to remove rough edges, and a knife to ream. With a flat dresser proceed to work the lead until it fits snugly into the brass ferrule (assume this type of ferrule is used). Approximately 1⅛″ (29 mm) to 1¼″ (32 mm) should be inserted into the ferrule. Next, using the shave hook, shave a portion equal to 1″ (25 mm) plus the insertion, say, 2¼″ (57 mm), and apply mutton tallow. Then insert into the ferrule and proceed to secure. Apply a ring of plumber's soil inside the ferrule brass side where lead ends inside.

Next, use dividers and mark a 1″ (25 mm) line around the lead from the face of the ferrule; next, use a wire brush to clean an additional 3″ (76 mm) of lead. Apply soil on this portion; try not to apply to the 1″ (25 mm) portion being readied to receive solder. Next, use a shave hook to clean a 1″ (25 mm) portion. Then reapply mutton tallow. Last, apply gummed paper to the

Fig. 10-5. Excess solder has been removed from the soiling and shaping has been started across the bottom and up the side next to the wiper.

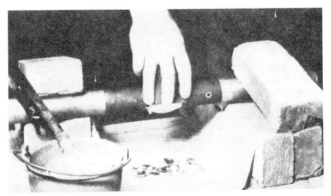

Fig. 10-6. The hand has been reversed and the stroke continues across the top and down the side away from the wiper.

ferrule. After wiping, prepare the lead cap, readying the stub for testing. A completed installation is shown in Fig. 10-8.

Lead Joining Work

In the regular run of lead-joining work, the plumber usually needs a pair of two-pound (907-gram) soldering irons. For ordinary work, the point of the soldering iron should be tinned on all four sides for a distance of at least ³⁄₄″ (19 mm) from the tip. If, however, soldering is to be done from underneath the joint, the iron should be tinned on one surface only — the side to be used next to the joint being soldered. This permits control of the solder and prevents it from running away from the joint.

Lap Joints

A lap of about ³⁄₈″ (10 mm) is advisable for the weights of lead ordinarily employed by plumbers. On the top surface of the bottom sheet, a line should be marked ¹⁄₂″ (12.7 mm) back from the edge to be joined. This portion should be shaved lightly with the shave hook, using strokes parallel to the edge, until the surface is clean back to the line.

The same procedure should be followed on the under side of the top sheet. The edge of the top sheet should also be cleaned and the top sheet placed in position, lapping over the other ³⁄₈″ (10 mm). This leaves ¹⁄₈″ (3.2 mm) of the cleaned portion of the lower piece of lead exposed. With the flat dresser, the top sheet should be dressed down to the level of the bottom sheet, except where it actually laps and is held up by the lead underneath. The lap should be dressed to fit snugly. With the shave hook, the upper surface of the top layer should then be cleaned for a distance of ³⁄₄″ (10 mm) from the edge. Tallow should be applied immediately to all cleaned areas. As in making butt joints, the sheets should be tacked in similar manner.

Fig. 10-7. A handy method of retaining under adverse conditions by building up the solder a little way from each side of the joint to be lifted off after completion.

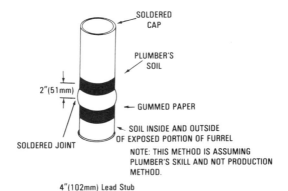

SOLDERED CAP

PLUMBER'S SOIL

2"(51mm)

GUMMED PAPER

SOIL INSIDE AND OUTSIDE OF EXPOSED PORTION OF FURREL

NOTE: THIS METHOD IS ASSUMING PLUMBER'S SKILL AND NOT PRODUCTION METHOD.

SOLDERED JOINT

4"(102mm) Lead Stub

Fig. 10-8. Lead stub wiped on a job brass ferrule.

Butt Joints

To make a butt joint, the edges to be joined should be beveled with a shave hook so that they make an angle of 45° or more with the vertical. This is accomplished by wrapping a piece of cloth around the index finger of the hand holding the shave hook and using this finger, pressed against the edge of the lead, as a guide when drawing the shave hook along the edge.

Immediately after shaving, tallow candle or refined mutton tallow free from salt should be rubbed over all shaved parts in a very thin coat to prevent oxidation.

Edges to be joined should then be placed firmly together and powdered rosin sprinkled along the joint. With a clean, well-tinned soldering iron and 50-50 solder, the edges are next tacked together at intervals of from 4″ (102 mm) to 6″ (152 mm), using a drop of solder at each point. An iron at proper heat should then be placed against the lead at the end of the seam in the groove formed by the abutting beveled edges.

Solder should be fed in slowly, allowing it to be melted by the iron and fill the groove. The iron should be drawn slowly along the joint, the speed being such as to permit the solder to melt and fill the groove continuously, building up to a slightly rounded surface when finished.

11
Lead and Oakum Joints

To prepare lead and oakum joints in pipe up to 6″ (152 mm), the following tools are needed:

1. 12 oz. or .340 kilogram ballpeen hammer.
2. Caulking iron(s).
3. Packing iron.
4. Yarning iron.
5. Joint runner (for pouring lead in horizontal joints).
6. Wood chisel (steel handle) for cutting "lead gate" created by use of joint runner.
7. Chain snap cutters or ratchet cutters (for cutting pipe).

A cold chisel and a 16-ounce (.453 kg.) ballpeen hammer can be used to cut pipe. Place a 2″ × 4″ or 5 cm × 10 cm piece of wood directly under the cut mark. Allow one end of the pipe to touch the ground, and with one foot placed near the desired cut, hold the pipe solidly against the 2″ × 4″.

Begin marking the pipe with light hammer blows. When the entire circumference of pipe is marked with cold chisel indentation marks, begin using heavier blows.

Since pipe length is 5′, a six-foot rule is usually called for in measuring. After the proper amount of oakum is placed in the joint, leaving 1″ (25 mm) for lead (joints up to 6″ or 152 mm), lead is then poured. The joint is now ready for caulking to make it water and air tight (Fig. 11-1).

Note: Before dipping the ladle into the molten lead, be sure it is dry and free from moisture. Warm it over the lead pot while the lead is being heated. Moisture on a ladle will form steam when dipped into molten lead, causing an explosion.

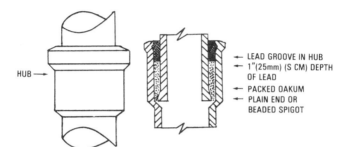

Fig. 11-1. Typical lead and oakum joint.

Table 11-1. Lead and Oakum Information

Size of Pipe and Fitting (Inch Joint)	Lead Ring Depth (Inches)	White Oakum (Ounces)	Brown Oakum (Ounces)	Lead SV-X (Lbs.)
2	1	1½	1¾	1¼
3	1	1¾	2½	1¾
4	1	2¼	3	2¼
5	1	2½	3¼	2¾
6	1	2¾	3½	3
8	1¼	5¼	7	6
10	1¼	6½	8½	8
12	1¼	7½	9¾	10¼
15	1½	11½	15	17¼

Table 11-2. Lead and Oakum Information (Metric)

Size of Pipe and Fitting	Lead Ring Depth	White Oakum	Brown Oakum	Lead SV-Xh
51 mm	25 mm	43 g	50 g	567 g
76 mm	25 mm	50 g	71 g	794 g
102 mm	25 mm	64 g	85 g	1020 g
127 mm	25 mm	71 g	92 g	1247 g
152 mm	25 mm	78 g	99 g	1360 g
203 mm	32 mm	149 g	198 g	2721 g
254 mm	32 mm	184 g	241 g	3629 g
305 mm	32 mm	213 g	276 g	4876 g
381 mm	38 mm	326 g	425 g	8051 g

Caulk horizontal joints inside first. Outside first is preferred on vertical joints. When caulking, use moderate hammer blows. Each position of the caulking iron should slightly overlap the previous position. The lead and oakum joint provides a waterproof joint: strong, flexible, root proof, water and air tight.

Note: Approximately 8 lbs. (3.62 kg) brown oakum is used per 100 lbs. (45.3 kg) of lead.

Six pounds (2.7 kg) white oakum used per 100 lbs. (45.3 kg) of lead.

Caulking lead in cast iron bell and spigot water mains should be 2 inches (51 mm) deep.

12
Silver Brazing and Soft Soldering

Table 12-1. Brazing Information

Size of Copper Tube		Oxygen Pressure		Acetylene Pressure	
Inches	Metric	psi	kPa	psi	kPa
½", ¾"	13 mm, 19 mm	5	34.5	5	34.5
1", 1¼"	25 mm, 32 mm	6	41.4	6	41.4
1½", 2", 2½"	38 mm, 51 mm, 64 mm	7	48.26	7	48.26
3", 3½"	76 mm, 89 mm	7½	51.7	7½	51.7
4", 5", 6"	102 mm, 127 mm, 152 mm	9	62	9	62

Applying Heat and Brazing Alloy

The preferred method is by the oxyacetylene flame. Propane and other gases are sometimes used on smaller tube sizes.

A slightly reducing flame should be used, with a slight feather on the inner blue cone; the outer portion of the flame is pale green. Heat the tube first, beginning at about one inch from the edge of the fitting. Sweep the flame around the tube in short strokes up and down at right angles to the run of the tube. It is very important that the flame be in continuous motion, and not be allowed to remain on any one point, to avoid burning through the tube.

Generally the flux may be used as a guide as to how long to heat the tube; continue heating after the flux starts to bubble or work and until the flux becomes quiet and transparent, like clear water.

Flux passes through four stages:

1. At 212°F (100°C) the water boils off.
2. At 600°F (316°C) the flux becomes white and slightly puffy and starts to work.
3. At 800°F (427°C) it lies against the surface and has a milky appearance.
4. At 1100°F (593°C) it is completely clear and active and has the appearance of water.

Avoid applying excess flux, and avoid getting flux on areas not cleaned. Particularly avoid getting flux into the inside of the tube itself. The purpose of flux is to dissolve residual traces of oxides, to prevent oxides from forming during heating, and to float out oxides ahead of the solder.

Now, switch the flame to the fitting at the base of the cup. Heat uniformly, sweeping the flame from fitting to tube until the flux on the fitting becomes quiet. Particularly avoid excessive heating of cast fittings.

When the flux becomes liquid and transparent on both the tube and the fitting, start sweeping the flame back and forth along the axis of the joint to maintain heat on the parts to be joined, especially toward the base of the cup of the fitting. The flame must be kept moving to avoid burning the tube or fitting.

When the joint has reached proper temperature, apply brazing wire or rod where the pipe enters the fitting. Keep the flame away from the rod or wire as it is fed into joint. Keep both the fitting and the tube heated by moving the flame back and forth from one to the other as the alloy is drawn into the joint.

When the joint is filled, a continuous fillet of brazing alloy will be visible completely around the joint. Stop feeding as soon as the joint is filled. *Note:* For larger size tube, 1″ (25 mm) and above, it is difficult to bring the whole joint up to heat at one time.

If difficulty is encountered in getting the entire joint up to the desired temperature, a portion of the joint can be heated and brazed at a time. At the proper brazing temperature, the alloy is fed into the joint and the torch is then moved to an adjacent area and the operation carried on progressively all around the joint, taking care to overlap each operation.

Horizontal and Vertical Joints

When making horizontal joints it is preferable to start applying the brazing alloy at the top, then the two sides, and finally the bottom, making sure that the operations overlap.

On vertical joints, it is immaterial where the start is made. If the opening of the socket is pointed down, care should be taken to avoid overheating the tube, as this may cause the alloy to run down the tube. If this condition is encountered, take the heat away and allow the alloy to set; then reheat the bank of the fitting to draw up the alloy. After the brazing alloy has set, clean off the remaining flux with a wet brush or swab. Wrought fittings may be chilled quickly. However, it is advisable to allow cast fittings to cool naturally to some extent before applying a swab.

If the brazing alloy refuses to enter the joint and tends to flow over the outside of either member of the joint, it indicates this member is overheated, or the other is underheated, or both. If the alloy fails to flow, or has a tendency to ball up, it indicates oxidation on the metal surfaces or insufficient heat on the parts to be joined.

Making Up a Joint

The preliminary steps of tube measuring, cutting, burr removing, and cleaning (tube ends and sockets must be thoroughly cleaned before beginning the brazing operation) are identical to the same steps in the soft-soldering process. A flux can be made that will be suitable for making silver solder joints on copper tubing by mixing powdered borax and alcohol or water to a thin, milky solution.

Soft Soldering

1. Avoid pointing the flame into the socket opening of the fitting.
2. Never apply the flame directly to solder.
3. On tubing 1" (25 mm) and above, a mild heating of the tube before playing the flame on the fitting is recommended. This ensures a better-made joint, assuring that solder is drawn into the joint by the natural force of capillary attraction.
4. Flame should be played at the base of the fitting, with the flame pointing in the direction of the socket opening. This ensures that any impurities, including excess flux, will be flushed out ahead of the solder as the joint is filled.

   If the flame is pointed toward the base of the fitting, there is a chance of these impurities or flux being trapped inside the joint, creating a flux pocket. A flux pocket prevents solder from completely occupying the inside of the socket.
5. When the material is hot enough, the flame should be moved away and the solder applied.
6. On larger-size tubing it is best to hold the flame long enough at the base of the fitting, then move it around the circumference, to ensure evenly distributed heat and solder.
7. As the joint cools, continue to apply solder around the entire face of the fitting; this will create a fillet that ensures a full joint. Heat rises and sometimes the top part of a horizontal

joint is too hot to retain solder, allowing it to run out; or in a tight, inaccessible place a portion may not have been heated. This can be detected by running solder around the entire face. If it runs smoothly around, creating a fillet, a better guarantee is received that a good joint has been made.

8. On horizontal joints it is recommended that the flame be played at the bottom of the fitting and solder applied at the top; however, this is merely preferred by the majority of plumbers.

Illustrations on Brazing and Soldering

Fig. 12-1. Fluxing. *(Courtesy Copper Development Association)*

Fig. 12-2. Assembling. *(Courtesy Copper Development Association)*

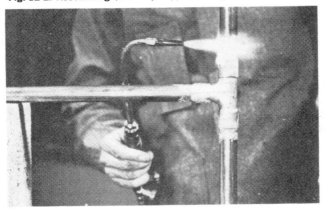

Fig. 12-3. Heating tube. *(Courtesy Copper Development Association)*

Fig. 12-4. Heating large tube. *(Courtesy Copper Development Association)*

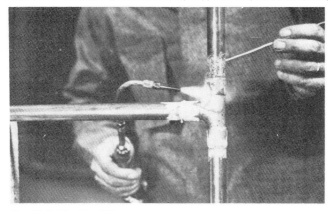

Fig. 12-5. Heating fitting. *(Courtesy Copper Development Association)*

Fig. 12-6. Heating large fitting. *(Courtesy Copper Development Association)*

Fig. 12-7. Feeding brazing alloy. *(Courtesy Copper Development Association)*

Fig. 12-8. Feeding upward. *(Courtesy Copper Development Association)*

Fig. 12-9. Swabbing. *(Courtesy Copper Development Association)*

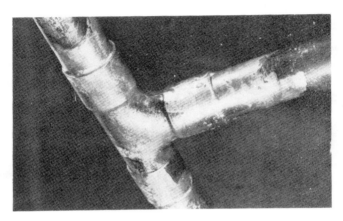

Fig. 12-10. Completed joint. *(Courtesy Copper Development Association)*

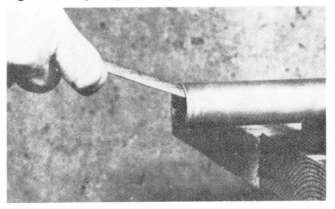

Fig. 12-11. Removing burrs. *(Courtesy Copper Development Association)*

Fig. 12-12. Cleaning tube end. *(Courtesy Copper Development Association)*

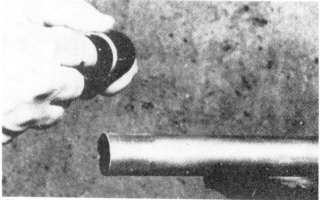

Fig. 12-13. Cleaning fitting socket. *(Courtesy Copper Development Association)*

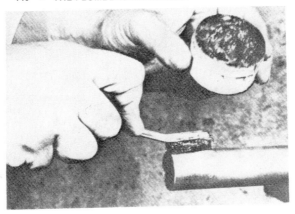

Fig. 12-14. Fluxing tube end. *(Courtesy Copper Development Association)*

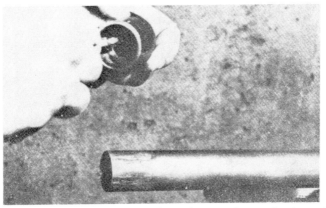

Fig. 12-15. Fluxing fitting socket. *(Courtesy Copper Development Association)*

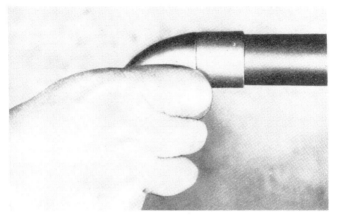

Fig. 12-16. Assembling. *(Courtesy Copper Development Association)*

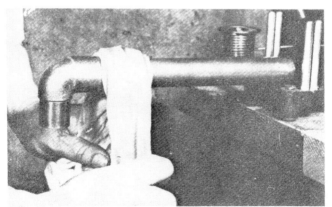

Fig. 12-17. Removing surplus flux. *(Courtesy Copper Development Association)*

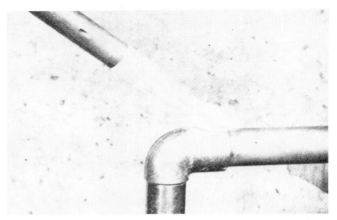

Fig. 12-18. Heating. *(Courtesy Copper Development Association)*

Fig. 12-19. Applying solder. *(Courtesy Copper Development Association)*

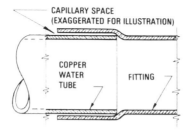

Fig. 12-20. A cross-sectional view of a joint positioned for soldering.

Basic Oxyacetylene Safety Measures

Safe practices to follow when oxyacetylene welding or cutting:

1. Always blow out cylinder valves before attaching the regulators. Dust can cause combustion, resulting in an explosion.

2. Stand to the side of the regulator when opening the cylinder valve. The weakest point of every regulator is either front or back. The regulator could blow out, and an explosion could occur.

3. Always release the adjusting screw on the regulator before opening the oxygen cylinder valve. When the adjusting screw is released, the seat of the regulator is in contact with the nozzle with sufficient pressure to hold the 2200 psi (15,168 kPa), so the oxygen released travels only a short distance. If regulator were open when high pressure is released through the seat nozzle, there would be expansion going into the regulator, then restriction into the nozzle, thus generating a lot of heat that could set off dust or oil.

4. Always open the cylinder valve slowly. By opening the valve slowly, the heat made from the travel is very small. The

main reason is to reduce shock. If the valve is opened fast, the pressure exerted from the shock hitting the seat surface exceeds that of the pressure contained in the cylinder.

5. A good practice is to light fuel gas before opening the oxygen valve on the torch. Light the large tip, show soot, then no soot, then flame leaving the tip. The burning rate should be set with acetylene valve only. If you do open the oxygen valve first, you pop the large tip.

6. Never use oil on regulators, torches, etc. Oxygen and oil create an explosion. In oxygen cylinders there is as much as 2200 psi (15,168 kPa) pressure. When the pressure is released from the cylinder through the regulator, the speed at which the oxygen travels exceeds the speed of sound and this generates heat and friction. The smallest amount of oil, just the oil from your skin, will ignite and blow up the regulator.

7. Do not store cylinders near flammable material, especially oil, grease, or any other readily combustible substance.

8. Acetylene cylinders should be stored in a dry, well-ventilated location.

9. Acetylene cylinders should not be stored in close proximity to oxygen cylinders.

10. Never tamper with safety devices in valves or cylinders. Keep sparks and flames away from acetylene cylinders and under no circumstances allow a torch flame to come in contact with safety devices. Should the valve outlet of an acetylene cylinder become clogged by ice, thaw with warm, not boiling, water.

11. Acetylene should never be used at a pressure exceeding 15 psi (103 kPa) gage.

12. The wrench used for opening the cylinder valve should always be kept on the valve spindle when the cylinder is in use.

13. Finally, points of suspected leakage should be tested by covering them with soapy water. *Never* test for leaks with an open flame.

PART II

Plumbing Systems

13
Plastic Pipe and Fittings

Plastic DWV piping has been approved by local and state codes including the Building Officials Conference of America, Southern Building Code Congress, International Association of Plumbing and Mechanical Officials, and FHA.

PVC: Type 1 polyvinyl chloride is strong, rigid, and economical. It resists a wide range of acids and bases, but may be damaged by some solvents and chlorinated hydrocarbons. The maximum service temperature is 140°F (60°C). PVC is better suited to pressure piping.

ABS: Usage of ABS has almost doubled compared with PVC in DWV piping systems; however, it is limited to 160°F (71.1°C) water temperatures at lower pressures considered adequate for DWV use.

CPVC: Chlorinated polyvinyl chloride meets national standards for piping 180°F (82.2°C) water at pressures of up to 100 psi (689 kPa). It can withstand 200°F (93.3°C) water temperature for limited periods. Chlorinated polyvinyl chloride is similar to PVC in strength and overall chemical resistance.

PE: Polyethylene is a flexible pipe for pressure systems. Like PVC, it cannot be used for hot water systems.

PB: Polybutylene is flexible and can be used for either hot or cold water pressure systems. Since no method has been found to chemically bond PB, solvent-weld joints *cannot* be used. Compression-type joints are used instead.

Polypropylene: This is a very lightweight material suitable for lower pressure applications up to 180°F (82.2°C). It is used widely for industrial and laboratory drainage acids, bases, and many solvents.

Kem-Temp (PVDF) or polyvinylidene fluoride: This is a strong, tough and abrasive-resistant fluorocarbon material. It has excellent chemical resistance to most acids, bases, and organic solvents and is ideally suited for handling wet or dry chloride, bromine, and other halogens. It can be used in temperatures of up to 280°F (138°C).

FRP epoxy: This is a fiberglass-reinforced thermoset plastic with high strength and good chemical resistance up to 220°F (104.4°C).

Expansion in Plastic Piping

PVC-Type 1: 100 feet or 30.5 meters operating at 140°F (60°C) will expand approximately 2″ or 5 cm.

CPVC: Polypropylene and PVDF at the same temperature would expand approximately 3¼″ or 8 cm.

Applications

Plastic pressure piping for hot and cold water supply is now permitted in FHA-financed rehabilitation projects. Plastic pipe enjoys markets in natural gas distribution, rural potable water systems, crop irrigation, and chemical processing. Almost 100% of all mobile homes and travel trailers have plastic pipe.

Two types of plastic pipe and fittings are commonly used for drainage systems: PVC and ABS.

Abbreviations

ABS	Acrylonitrile Butadiene Styrene
PVC	Polyvinyl Chloride
NSF	National Sanitation Foundation
PPI	Plastic Pipe Institute
CPVC	Chlorinated Polyvinyl Chloride
PE	Polyethylene Plastic or Resin
PVDF	Polyvinylidene Fluoride
PB	Polybutylene

Joints with Plastic Tubing

Cutting the Tubing

It is important to use the right primer and/or solvent. Priming is essential with PVC and CPVC. No priming is needed with ABS. The *recommended practice* for making solvent-cemented joints with PVC and ABS pipe and fittings follows. Pipe should be cut square, using a fine-tooth hand saw and a miter box, or a fine-tooth power saw with a suitable guide (Fig. 13-1). Regular pipe cutters may also be used (a special cutting wheel is available to fit standard cutters).

Great care should be taken to remove all burrs and ridges raised at the pipe end (Fig. 13-2). If the ridge is not removed, cement in the fitting socket will be scraped from the surface on insertion, producing a dry joint and causing probable joint

Table 13-1. PVC Water Pressure Ratings at 73.4° (23°C) for Schedule 40

No. PVC-1120-B PVC-1220-B Pipe PVC-2120-B Size CPVC-4120-B	PVC-2110-B		PVC-2112-B		PVC-2116-B		CPVC-4116-B	
	psi	kPa	psi	kPa	psi	kPa	psi	kPa
⅜" (9.5 mm)	620	4275	310	2137	390	2689	500	3447
½" (12.7 mm)	600	4137	300	2068	370	2551	480	3310
¾" (19 mm)	480	3310	240	1655	300	2068	390	2689
1" (25.4 mm)	450	3103	220	1517	280	1931	360	2482
1¼" (31.75 mm)	370	2551	180	1241	230	1586	290	2000
1½" (38 mm)	330	2275	170	1172	210	1448	260	1793
2" (51 mm)	280	1931	140	965	170	1172	220	1517
2½" (63.5 mm)	300	2068	150	1034	190	1310	240	1655
3" (76 mm)	260	1793	130	896	160	1103	210	1448
3½" (89 mm)	240	1655	120	827	150	1034	190	1310
4" (101.6 mm)	220	1517	110	758	140	965	180	1241
5" (127 mm)	190	1310	100	689	120	827	160	1103
6" (152.4 mm)	180	1241	90	621	110	758	140	965

Table 13-2. ABS Water Pressure Ratings at 73.4°F (23°C) for Schedule 40

Nominal Pipe Size	ABS-1210		ABS-1316		ABS-2112	
	psi	kPa	psi	kPa	psi	kPa
½" (12.7 mm)	298	2055	476	3282	372	2465
¾" (19 mm)	241	1662	385	2655	305	2103
1" (25.4 mm)	225	1551	360	2482	282	1944
1¼" (31.75 mm)	184	1269	294	2027	229	1579
1½" (38 mm)	165	1138	264	1820	207	1427
2" (51 mm)	139	958	222	1531	173	1193
2½" (63.5 mm)	152	1048	243	1675	190	1310
3" (76 mm)	132	910	211	1455	165	1138
4" (101.6 mm)	111	765	177	1220	138	951
6" (152.4 mm)	88	607	141	972	110	758

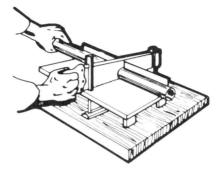

Fig. 13-1. Cut the pipe squarely. *(Courtesy NIBCO, Inc.)*

failure. All burrs should be removed with a knife, file, or abrasive paper.

Test Fit the Joint

Wipe both the outside of the pipe and the socket of the fitting with a clean, dry cloth to remove foreign matter. Mate the two parts without forcing. Measure and mark the socket depth of the fitting on the outside of the pipe. Do not scratch or damage pipe surface to indicate when the pipe end will be bottomed.

The pipe should enter the fitting at least one-third but not more than half way into the socket or fitting depth (Fig. 13-3). If the pipe will not enter the socket by that amount, the diameter may be reduced by sanding or filing. Extreme care should be taken not to gouge or flatten the pipe end when reducing the diameter. Unlike copper fittings, the inside walls of plastic fitting sockets are tapered so that the pipe makes contact with the sides of the fitting wall before the pipe reaches the seat of the socket.

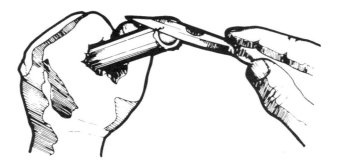

Fig. 13-2. Smooth the end of the pipe. *(Courtesy NIBCO, Inc.)*

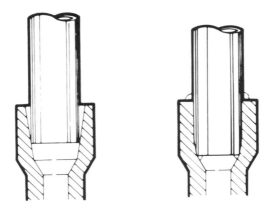

Fig. 13-3. Cross-sectional drawing of PVC joint. *(Courtesy NIBCO, Inc.)*

Surfaces to be joined should be clean and free of moisture
before application of the cement. The outside surface of the
pipe (for socket depth) and the mating socket surface are to be
cleaned and the gloss removed with the recommended chemical cleaner.

An equally acceptable substitute is to remove the gloss from
the mating surfaces (both pipe and socket) with abrasive paper
or steel wool. Wipe off all particles of abrasive and/or PVC
before applying primer or cement.

Apply Primer (PVC and CPVC only)

Use only primer formulated for PVC or CPVC. Apply first to
the inside of the fitting (Fig. 13-4), then to the outside of the
pipe to the depth that will be taken into the fitting when seated.
Wait 5 to 15 seconds before applying cement. A primer is not
needed for ABS (Fig. 13-5).

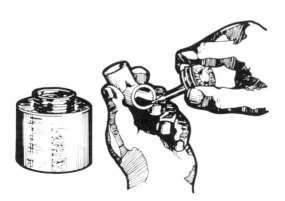

Fig. 13-4. Apply primer (PVC and CPVC only). *(Courtesy NIBCO, Inc.)*

Fig. 13-5. Apply solvent cement. (Courtesy NIBCO, Inc.)

Apply Cement

Handling cement: Keep the cement can closed and in a shady place when not in use. Discard the cement when an appreciable change in viscosity takes place, or when a gel condition is indicated by noting that cement will not flow freely from the brush, or when cement appears lumpy and stringy. The cement should not be thinned. Keep the brush immersed in cement between applications.

Apply the solvent cement while the surfaces are still wet from the primer. Brush on a full, even coating to the inside of the fitting. Be careful not to form a puddle in the bottom of the fitting. Next, apply solvent to the pipe to the same depth as you applied primer. Recoat the pipe with a second uniform coat of cement, including the cut end of the pipe.

The cement is applied with a natural bristle or nylon brush, using a 1/2" (13 mm) brush for nominal pipe sizes 1/2" (13 mm) and less, 1" (24 mm) wide brush for pipe up through 2" (51 mm) nominal size, and a brush width at least 1/2 of nominal pipe size for sizes above 2″ (51 mm), except that for pipe sizes 6″ (152 mm) and larger a 2½″ (64 mm) brush is adequate.

Special Instructions for Bell End Pipe

The procedure as stated thus far may be followed in the case of bell end pipe, except that great care should be taken not to apply an excess of cement in the bell socket, nor should any cement be applied on the bell-to-pipe transition area. This precaution is particularly important for installation of bell end pipe with a wall thickness of less than ⅛″ (3 mm).

Assembly of Joint

Solvent cement dries quickly. Work fast. Immediately after applying the last coat of cement to the pipe, insert the pipe into the fitting until it bottoms at the fitting shoulder. Turn the pipe or fitting ¼ turn during assembly (but not after the pipe is bottomed) to evenly distribute the cement. Assembly should be completed within 20 seconds (longer in cold weather) after the last application of cement. The pipe should be inserted with a steady even motion. Hammer blows should not be used (Fig. 13-6).

Until the cement is set in the joint, the pipe may back out of the fitting socket if not held in place for approximately one minute after assembly.

Care should be taken during assembly not to disturb or apply any force to joints previously made. Fresh joints can be destroyed by early rough handling.

After assembly, wipe excess cement (Fig. 13-7) from the pipe at the end of the fitting socket. A properly made joint will normally show a bead (Fig. 13-8) around its entire perimeter. Any gaps in the bead may indicate a defective assembly job, due to insufficient cement or use of light-bodied cement on a gap fit where heavy-bodied cement should have been used.

Fig. 13-6. Fit and position pipe and fitting. *(Courtesy NIBCO, Inc.)*

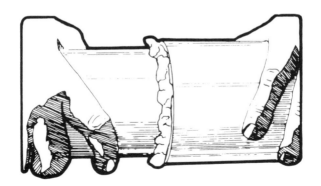

Fig. 13-7. Check for the correct bead. *(Courtesy NIBCO, Inc.)*

Set Time

Handle the newly assembled joints carefully until the cement has gone through the set period. Recommended set time is related to temperature.

Temperature	Set Time
60°F to 100°F (15.5 to 37.7°C)	30 minutes
40°F to 60°F (4.4 to 15.5°C)	1 hour
20°F to 40°F (–6.6 to 4.4°C)	2 hours
0°F to 20°F (–17.7 to –6.6°C)	4 hours

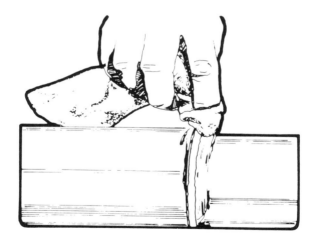

Fig. 13-8. Wipe off excess cement. *(Courtesy NIBCO, Inc.)*

Plastic Pipe and Fittings

After the set period, the pipe can be carefully placed in position. If pipe is to be buried, shade backfill, leaving all joints exposed so that they can be examined during pressure tests.

Test pressure should be 150% of system design pressure and held at this pressure until the system is checked for leaks, or follow the requirements of the applicable code, whichever is higher.

Note: For most cases, 48 hours is considered to be a safe period for the piping system to be allowed to stand vented to the atmosphere before pressure testing. Shorter periods may be satisfactory for high air temperatures, small sizes of pipe, quick drying cement, and tight dry fit joints.

PVC and ABS pipe and fittings may be stored either inside or outdoors if they are protected from direct sunlight. The plastic pipe should be stored in such a manner as to prevent sagging or bending.

Plastic pipe should be supported in horizontal runs as follows:

Nominal Pipe Size	Schedule 40
$\frac{1}{2}$" and $\frac{3}{4}$" (12.7 and 19 mm)	Every 4 ft. (1.22 m)
1" and $1\frac{1}{4}$" (25 and 32 mm)	Every $4\frac{1}{2}$ ft. (1.37 m)
$1\frac{1}{2}$" and 2" (38 and 51 mm)	Every 5 ft. (1.52 m)
3" (76 mm)	Every 6 ft. (1.83 m)
4" (102 mm)	Every $6\frac{1}{4}$ ft. (1.9 m)
6" (152 mm)	Every $6\frac{3}{4}$ ft. (2.06 m)

Nominal Pipe Size	Schedule 80
$\frac{1}{2}$" and $\frac{3}{4}$" (12.7 and 19 mm)	Every 5 ft. (1.52 m)
1" and $1\frac{1}{4}$" (25 and 32 mm)	Every $5\frac{1}{2}$ ft. (1.68 m)
$1\frac{1}{2}$" and 2" (38 and 51 mm)	Every 6 ft. (1.83 m)
3" (76 mm)	Every 7 ft. (2.13 m)
4" (102 mm)	Every $7\frac{1}{2}$ ft. (2.29 m)
6" (152 mm)	Every $8\frac{1}{2}$ ft. (2.59 m)

The industry does not recommend threading ABS or PVC Schedule 40 plastic pipe.

A quart can of the solvent recommended for the type of pipe being joined is generally sufficient for the average two-bath home.

An ABS stack can be tested within one hour after the last joint is made up.

Common pipe dopes must not be used on threaded joints. Some pipe lubricants contain compounds that may soften the surface, which, under compression, can set up internal stress corrosion. If a lubricant is believed necessary, ordinary Vaseline or pipe tape can be used.

Do not use alcohol or antifreeze solutions containing alcohol to protect trap-seals from freezing. Strong saline solutions or magnesium chloride in water (22% by weight) can be used safely. Glycerol (60% by weight) mixed with water is also recommended.

ABS absorbs heat so slowly that once installed, heat from dishwashers, clothes washers, and the discharge from similar installations will not cause any problem.

PVC solvent cements are available in two general viscosity categories: light and standard. The light cements are intended for use with pipes and fittings up through 2" (51 mm) N.P.S., and for pipes and fittings where "interface fits" occur between the parts to be joined.

Last, but not least, always look for the initials NSF; these initials stand for National Sanitation Foundation, denoting approval and standards met as handed down by NSF. The foundation is a nonprofit, noncommercial organization seeking solutions to all problems involving cleanliness. It is dedicated to the prevention of illness, the promotion of health, and the enrich-

NIBCO's new CPVC
Transition Union

Copper end
available in
a variety
of connections

Buna-N Gasket
for a tight
seal

CPVC Tail-
piece
solvent welds
direct to tube

Brass
take-up
nut tightens
securely

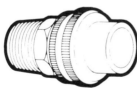

MPT to CPVC

FPT to CPVC

Fig. 13-9. NIBCO CPVC-to-metal unions and adapters. *(Courtesy NIBCO, Inc.)*

Sweat copper to CPVC

Compression to CPVC

Drop-Ear Ell
FPT to CPVC

Stop-and-Waste
Valve has union-
connected CPVC
tailpieces

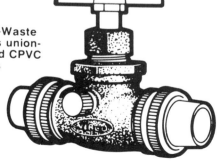

ment of the quality of American living through the improvement of the physical, biological, and social environment in which we live today. The NSF seal on DWV and potable water plastic pipe and fittings means compliance with the foundation's policies and standards.

The NSF standards, research, and education programs are designed to benefit all parties: the manufacturer, the regulatory officials, the building industries, the product specifier, the installer, and the ultimate user; but their most important function may be their aid in providing protection to public health.

Solvent weld joints between plastic and metals are impossible for obvious reasons. There is no way to achieve a satisfactory bond. When making a transition between plastic and metal, temperature is a critical factor. For cold water lines you simply use a threaded adapter. However, this method is not recommended with hot water. Metal and plastic expand and contract at different rates, working against each other in effect. Since a threaded adapter cannot compensate for this uneven expansion and contraction, the joint may eventually leak.

To solve this problem a variety of transition adapters are available. But some introduce serious flow restrictions. The CPVC-to-metal adapters shown in Fig. 13-9 were designed by NIBCO to avoid this problem.

Connecting CPVC to a Water Heater and Existing Plumbing

At least 12 inches (31 cm) of metal pipe is recommended on both hot and cold water lines above the tank (Fig. 13-10) before converting to CPVC. Transition unions are used to connect the CPVC to the metal pipe risers.

Codes generally require the installation of a combination temperature/pressure relief valve on the heater tank. CPVC tubing may be used for the relief line to a point 6" (51 cm) above the floor. The relief line must be the same size as the outlet of the relief valve.

CPVC tubing should be supported at least every 3' (91 cm) in runs longer than 3' (91 cm). Some allowances must be made for

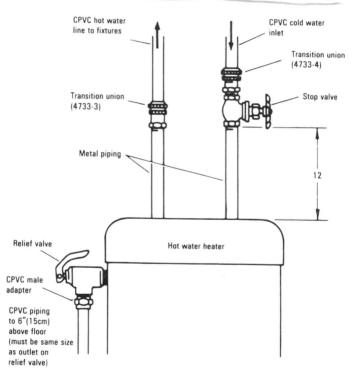

Fig. 13-10. Hot water heater connections with CPVC. *(Courtesy NIBCO, Inc.)*

expansion, especially on hot water lines. A 12″ (31 cm) offset every 10′ (3 m) in a straight run is recommended.

Connecting ABS and PVC Drainage Systems to Other Materials

For DWV systems, most codes require that special adapter fittings be used when making a transition from one material to another. NIBCO offers a full selection of adapters, from threaded adapters (since there is very little thermal expansion in most DWV lines), to no-hub adapters for connecting to no-hub pipe.

Special lead-joint plastic adapters (Fig. 13-11) for soil pipe are also offered. Plastic can be lead caulked into cast iron without damage, since the lead cools much faster than plastic absorbs heat. However, the lead should not be heated beyond normal.

Special note: Plastic pipe lighter than Schedule 80 should not be threaded. The wall thickness remaining after threading will not provide adequate strength.

Installing Multistory Stacks

Both ABS and PVC are used extensively in multistory installations. Although the thermal expansion and contraction rate of plastics is greater than that of metal, the *force* behind the expansion is not as great, so it is more easily controlled. For two-story buildings, there is no problem. For buildings three stories or more (Model Codes have approved plastic DWV for high-rise installations), use a restraining fitting (Fig. 13-12) every second floor. This fitting prevents vertical expansion, which could possibly damage the lines leading into the stack, but allows horizontal or circumferential expansion, to relieve any stress.

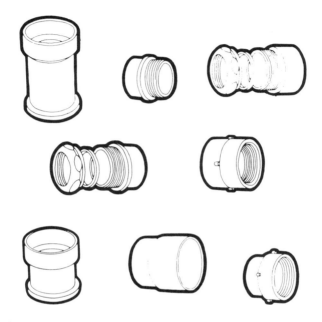

Fig. 13-11. Drainage system adapters. *(Courtesy NIBCO, Inc.)*

Supporting Plastic DWV Systems

Plastic DWV systems should be supported, just like any other type of system. The vertical stack should be supported, but remember to allow for horizontal expansion. A noise problem is likely to arise if you block or wedge the stack against wood framing without space for expansion. Likewise, branch fittings serving trap arms should be supported where necessary, but

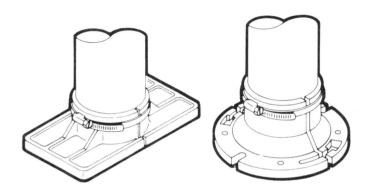

Fig. 13-12. Restraining fittings for multistory buildings. *(Courtesy NIBCO, Inc.)*

remember to allow for expansion. When suspended below the floor, hanger support is recommended every four or five feet.

Allowing for Expansion/Contraction in a Plastic DWV System

Both ABS and PVC have a coefficient of expansion higher than most metal systems. Normal installations, with relatively short runs (less than 50′ or 15.24 m), present little or no problem with proper support. The expansion and contraction will be contained between the supports, and the material will stress-relieve itself.

Where changes in direction are abrupt and followed by only a very short run, such as in a 45° or 90° offset, the support at the change of direction should be tightly clamped to the pipe. This avoids heavy flex loading of fittings at that point. Where changes in direction are followed by long, straight runs, supports next to the change in direction should be loose to allow movement of the pipe through the support.

Testing Time

The ABS stack can be tested one hour after the last joint is made. The warmer the ambient temperature, the faster the set.

PVC requires somewhat more time for initial set and curing. The accompanying chart gives you a guide for initial set time (from solvent application to "safe to handle" time), and a guide for joint cure time (from solvent application to testing), for PVC solvent welded systems.

Polybutylene Plumbing System

Polybutylene is said to offer a unique combination of flexibility, toughness, stress crack resistance, and creep resistance. Fittings (Fig. 13-13) are manufactured from Celcon® which is said to be impervious to scale buildup and corrosion. Certain fittings are also made from brass for special applications. Celcon® is a widely accepted material for use in potable plumbing systems.

Polybutylene pipe resists corrosion, freezing, rust, acidic soils, and scale buildup. It will not burst when frozen and withstands temperatures of 180°F (82.2°C) at 100 psi (689 kPa). Polybutylene is also unaffected by freezing. The pipe will expand to accommodate the ice. There is no splitting or cracking as with other materials.

It weighs less than 50 lbs. (22.68 kg) for 1000 feet (304.8 m) of $1/2''$ (12.7 mm) nominal tubing. Standard nominal sizes are $1/4$, $3/8$, $1/2$, $3/4$, 1, $1 1/4$, $1 1/2$, and 2 inch; or 6.4, 9.5, 12.7, 19, 25, 32, 38, and 51 mm. Pipe up to 6'' (15 cm) in diameter is available. Polybutylene is a poor conductor of heat, which means heat loss through pipes is reduced. With PB tubing, water hammer is virtually nonexistent, because PB is flexible.

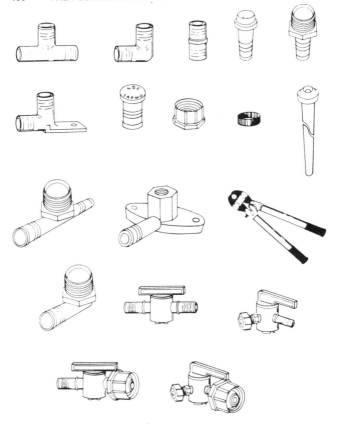

Fig. 13-13. Crimp fittings for PB plastic tubing. *(Courtesy Wrightway Mfg. Co.)*

Polybutylene pipe is recommended for these applications:

1. In-house plumbing for hot and cold water systems.
2. Commercial water supply.
3. Slab heat. (Most areas approve PB both above and in slab. Check your plumbing codes.)
4. Hydronic heat.
5. Solar (check with the manufacturer for specific applications.)

It is produced in rolls from 100' to 1000' (30.48 m to 304.8 m) and in 20' (6.1 m) straight lengths, in three colors: white, gray, and black. Polybutylene should be stored out of direct sunlight as it is susceptible to ultraviolet attack. Black PB has the best ultraviolet protection.

Crimped Joints

Polybutylene plumbing systems are designed for the professional and therefore require special tools. The crimping tool must be calibrated at least twice a day. On the Wrightway CTGAGE, there are two steps, a go and no-go section for each size crimping tool. One gage works for all three crimp dimensions. The smaller diameter is called the go section. To use, simply place the gage inside the tool and crimp around the go section. Then tighten or loosen the adjusting screw located near the top handle of the tool until the gage fits snugly. The gage should have some drag as it is turned inside the tool. Do not overtighten.

A loose tool will result in an insufficient crimp and a tool that is too tight may damage both the fitting and the tool. Remember to gage your tool before you start the day and during the day, depending upon its use. *Note:* Lubricate the tool at least once a week to get maximum life. Check to make sure there is no looseness in the tool piece parts and that all assembly bolts are tight when adjusting the tool.

Step-by-step crimping instructions (Fig. 13-14):

1. Cut the tubing squarely with a tubing cutter designed for plastic. The cutter should be rotated to slice into the tubing for a better cut.
2. Slip the crimp ring over the tubing.
3. Insert the fitting into the tube as far as possible. Position the ring over the center of the inserted fitting [approximately 1/4″ (6.4 mm) from the shoulder].
4. Set the crimping tool over the ring. Be sure the tool is square and centered on the ring. Close the tool until the stops are reached.
5. A properly crimped fitting should never leak or pull apart. A gage (QC43SP) to measure crimped connections is available from U.S. Brass. A 3/4″ (19.1 mm) connection should fit into the 0.960″ (2.44 cm) end and a 1/2″ (12.7 mm) connection should fit the 0.715″ (18.2 mm) end.

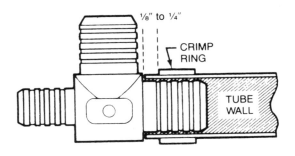

Fig. 13-14. Cross-section of PB crimped joint. *(Courtesy U.S. Brass)*

6. In the event of an improperly crimped fitting, remove the fitting and begin again at step 1.
7. On the threaded transition fittings, do not overtighten. It is recommended that Teflon® tape be used on all threaded connections.

Heat-Fusion Joints

The Wrightway heat-fusion tool (Fig. 13-15) must simply be plugged into any grounded 110-volt outlet. Approximately 15 minutes are needed for the tool to warm up. Care should be taken in securing the tool to avoid accidental burns.

The tool is designed with sockets and nipples that correspond to the tube size you are going to join. Take a piece of scrap tube and touch a socket or nipple. If the tube starts to soften and gives off a slight amount of smoke, your tool is ready for use. The smoke is nontoxic.

Fig. 13-15. Wrightway heat-fusion tool. *(Courtesy U.S. Brass)*

Place a brass ferrule inside of the tubing to be used. Its function is to keep the tube from collapsing during the fusion. Push a fitting on the fusion tool nipple that corresponds to the size of the fitting. Place the tube with the ferrule inside the corresponding socket. Count to three. Pull the tube out of the socket; then pull the fitting off the nipple. Push the tube into the socket until it stops. The joint is complete. The total operation should take less than 15 seconds.

What happened: As you slid the tube into the fitting, you mixed the two melted surfaces, forming one piece. Because of the increased wall thickness (tube plus fitting), your joint is now stronger than the tube itself.

A few tips:

1. Keep the tool clean. Use a fitting brush or steel wool.
2. Do not overheat. You are overheating if the tube or fitting loses its shape.
3. Do not join an overheated part. The tube will flare over outside of fitting.

Compression Fittings

An end-compression type of connection is shown in Fig. 13-16. It is a conventional-looking Celcon male fitting with threaded ends. The plumbing connection is made with a sealing cone, retaining ring, and nut placed on the tube. The nut-ring-cone assembly is simply threaded onto the fitting. *No special tools.* A pocket knife and pliers do the job. The nut is hand tightened, then given a turn or two with pliers or wrench. This connection is ideal for repair plumbing and for nonprofessional use.

Fig. 13-16. Qicktite® compression fitting. *(Courtesy U.S. Brass)*

Qest threaded fittings use standard plumbing threads. They can be connected directly to *any other* standard plumbing thread. Nut-ring-cone Qicktite components can be applied directly to copper and CPVC tube as well as to polybutylene.

Friction Welding

Friction welding is a method used by Qest in manufacturing supply tubes. The cone is spun onto the tube at 2500 rpm, creating friction and heat sufficient to weld the two pieces together. Portable friction-weld tools can be made available to professional plumbers and plumbing manufacturers.

14
Cast Iron Pipe and Fittings

Members of the Cast Iron Soil Pipe Institute adopted an insignia for use as a symbol of quality and to provide a simple method of designating the standard desired. At the present time, the CISPI members consist of 12 companies, operating 17 plants in 9 states.

Cast Iron Soil Pipe Institute Member Foundries

The American Brass & Iron
   Foundry
7825 San Leandro Street
Oakland, CA 94621
(415) 632-3467

American Foundry
P.O. Box 22045
Los Angeles, CA 90022
(213) 723-3695

Anaheim Foundry
800 East Orangethorpe Ave.
Anaheim, CA 92801
(714) 870-9000

Charlotte Pipe & Foundry
   Company
P.O. Box 35430
Charlotte, NC 28235
(704) 372-5030

The Eastern Foundry Company
Spring & Shaffer Streets
Boyertown, PA 19512
(215) 367-2153

Griffin Pipe Products
   Company
Home Office: 2000 Spring Rd.
Oakbrook, IL 60521
(312) 654-2500

Griffin Pipe Products
   Company
P.O. Drawer 740
Lynchburg, VA 24505
(804) 845-8021

Jones Manufacturing Co.
P.O. Box 6696
Birmingham, AL 35210
(205) 956-5511

Krupp Quakertown
Foundry, Inc.
4th and Mill Streets
Quakertown, PA 18951
(215) 538-2920

Southeastern Specialty &
Manufacturing Co.
Home Office:
930 South Southlake Drive
Hollywood, FL 33019
(305) 922-6736

Southeastern Specialty &
Manufacturing Co.
P.O. Box 2048
Anniston, AL 36202
(205) 237-6641

Tyler Pipe Industries
P.O. Box 2027
Tyler, TX 75710
(214) 882-5511

Tyler Pipe Industries
P.O. Box 35
Macungie, PA 18062
(215) 966-3491

U.S. Pipe & Foundry Co.
Home Office:
P.O. Box 10406
Birmingham, AL 35202
(205) 254-7261

U.S. Pipe & Foundry Co.
P.O. Box 788
Anniston, AL 36202
(205) 831-3660

U.S. Pipe & Foundry Co.
P.O. Box 6129
Chattanooga, TN 37401
(615) 265-4611

Universal Cast Iron
Manufacturing Co.
5404 Tweedy Place
South Gate, CA 90280
(213) 569-8151

Cast Iron Soil Pipe Fittings and Specifications

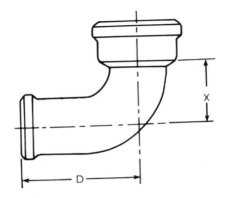

Fig. 14-1. ¼ bend.

Table 14-1. ¼ Bend

Size		D		X	
Inches	mm	Inches	mm	Inches	mm
2	51	6	152	3¼	83
3	76	7	178	4	102
4	102	8	203	4½	114
5	127	8½	216	5	127
6	152	9	229	5½	140
8	203	11½	292	6⅝	168
10	254	12½	318	7⅝	194

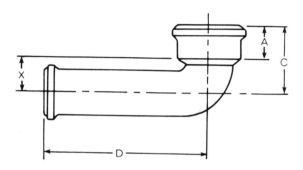

Fig. 14-2. Long low hub ¼ bends.

Table 14-2A. Long Low Hub ¼ Bends (Inches)

Size	D	A	X	C
4" × 12"	12"	3"	2¾"	5¾"
4" × 14"	14"	3"	2¾"	5¾"
4" × 16"	16"	3"	2¾"	5¾"
4" × 18"	18"	3"	2¾"	5¾"

Table 14-2B. Long Low Hub ¼ Bends (mm)

Size	D	A	X	C
102 × 305 mm	305 mm	76 mm	70 mm	146 mm
102 × 356 mm	356 mm	76 mm	70 mm	146 mm
102 × 406 mm	406 mm	76 mm	70 mm	146 mm
102 × 457 mm	457 mm	76 mm	70 mm	146 mm

Note: Where space is limited in reference to water closets, 1¾"
(44 mm) is gained in measurement C.

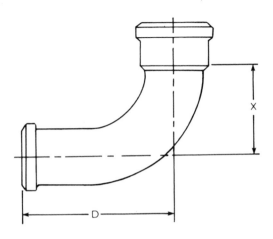

Fig. 14-3. Sweeps.

Table 14-3A. Short Sweeps

Size		D		X	
Inches	mm	Inches	mm	Inches	mm
2	51	8	203	5¼	133
3	76	9	229	6	152
4	102	10	254	6½	165
5	127	10½	267	7	178
6	152	11	279	7½	191
8	203	13½	343	8⅝	219
10	254	14½	368	9⅝	244

Table 14-3B. Long Sweeps

Size		D		X	
Inches	mm	Inches	mm	Inches	mm
2	51	11	279	8¼	210
3	76	12	305	9	229
4	102	13	330	9½	241
5	127	13½	343	10	254
6	152	14	356	10½	267
8	203	16½	419	11⅝	295
10	254	17½	445	12⅝	321

Table 14-3C. Reducing Long Sweeps

	Size	D	X
Inches	3 × 2	9	6
	4 × 3	10	6½
mm	76 × 51	229	152
	102 × 76	254	165

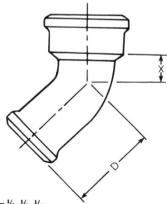

Fig. 14-4. Bends — ⅙, ⅛, 1/16.

Table 14-4A. $\frac{1}{8}$ Bends

Size		D		X	
Inches	mm	Inches	mm	Inches	mm
2	51	$4\frac{1}{4}$	108	$1\frac{1}{2}$	38
3	76	$4\frac{15}{16}$	125	$1\frac{15}{16}$	49
4	102	$5\frac{11}{16}$	144	$2\frac{3}{16}$	56
5	127	$5\frac{7}{8}$	149	$2\frac{3}{8}$	60
6	152	$6\frac{1}{16}$	154	$2\frac{9}{16}$	65
8	203	8	203	$3\frac{1}{8}$	79
10	254	$8\frac{3}{8}$	213	$3\frac{1}{2}$	89

Table 14-4B. $\frac{1}{16}$ Bends

Size		D		X	
Inches	mm	Inches	mm	Inches	mm
2	51	$3\frac{5}{8}$	92	$\frac{7}{8}$	22
3	76	$4\frac{3}{16}$	106	$1\frac{3}{16}$	30
4	102	$4\frac{13}{16}$	122	$1\frac{5}{16}$	33
5	127	$4\frac{7}{8}$	124	$1\frac{3}{8}$	35
6	152	5	127	$1\frac{1}{2}$	38
8	203	$6\frac{11}{16}$	170	$1\frac{13}{16}$	46
10	254	$6\frac{7}{8}$	175	2	51

Table 14-4C. $\frac{1}{6}$ Bends

Size		D		X	
Inches	mm	Inches	mm	Inches	mm
2	51	$4\frac{3}{4}$	121	2	51
3	76	$5\frac{1}{2}$	140	$2\frac{1}{2}$	64
4	102	$6\frac{5}{16}$	160	$2\frac{13}{16}$	71
5	127	$6\frac{5}{8}$	168	$3\frac{1}{8}$	79
6	152	$6\frac{7}{8}$	175	$3\frac{3}{8}$	86
8	203	9	229	$4\frac{1}{8}$	105
10	254	$9\frac{9}{16}$	243	$4\frac{11}{16}$	119

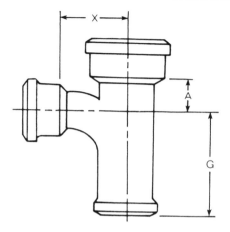

Fig. 14-5A. Single and double sanitary T-branches.

Table 14-5A. Single and Double Sanitary T-Branches (Inches)

Size	X	G	A
2	2¾	6¼	1¾
3	4	7½	2½
4	4½	8	3
5	5	8½	3½
6	5½	9	4
3 × 2	4	7	2
4 × 2	4½	7	2
4 × 3	4½	7½	2½
5 × 2	5	7	2
5 × 3	5	7½	2½
5 × 4	5	8	3
6 × 2	5½	7	2
6 × 3	5½	7½	2½
6 × 4	5½	8	3
6 × 5	5½	8½	3½

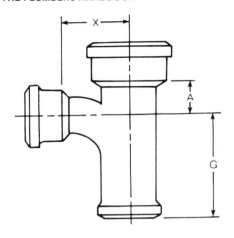

Fig. 14-5B. Single and double sanitary T-branches.

Table 14-5B. Single and Double Sanitary T-Branches (mm)

Size	X	G	A
51	70	159	44
76	102	191	64
102	114	203	76
127	127	216	89
152	140	229	102
76 × 51	102	178	51
102 × 51	114	178	51
102 × 76	114	191	64
127 × 51	127	178	51
127 × 76	127	191	64
127 × 102	127	203	76
152 × 51	140	178	51
152 × 76	140	191	64
152 × 102	140	203	76
152 × 127	140	216	89

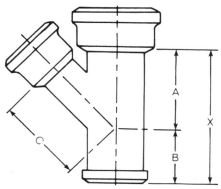

Fig. 14-6A. Y-branches, single and double.

Table 14-6A. Y-Branches, Single and Double (Inches)

Size	A	B	C	X
2	4	4	4	8
3	5½	5	5½	10½
4	6¾	5¼	6¾	12
5	8	5½	8	13½
6	9¼	5¾	9¼	15
8	11¹³⁄₁₆	7¹¹⁄₁₆	11¹³⁄₁₆	19½
10	14½	8	14½	22½
3 × 2	4¹³⁄₁₆	4³⁄₁₆	5	9
4 × 2	5⅜	3⅝	5¾	9
4 × 3	6¹⁄₁₆	4⁷⁄₁₆	6¼	10½
5 × 2	5⅞	3⅛	6½	9
5 × 3	6⅝	3⅞	7	10½
5 × 4	7⁵⁄₁₆	4¹¹⁄₁₆	7½	12
6 × 2	6⁷⁄₁₆	2⁹⁄₁₆	7¼	9
6 × 3	7⅛	3⅜	7¾	10½
6 × 4	7¹³⁄₁₆	4³⁄₁₆	8¼	12
6 × 5	8⁹⁄₁₆	4¹⁵⁄₁₆	8¾	13½

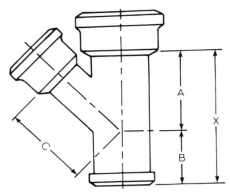

Fig. 14-6B. Y-branches, single and double.

Table 14-6B. Y-Branches, Single and Double (mm)

Size	A	B	C	X
51	102	102	102	203
76	140	127	140	267
102	171	133	171	305
127	203	140	203	343
152	235	146	235	381
203	300	195	300	495
254	368	203	368	572
76 × 51	122	106	127	229
102 × 51	137	92	146	229
102 × 76	154	113	159	267
127 × 51	149	79	165	229
127 × 76	168	98	178	267
127 × 102	186	119	191	305
152 × 51	164	65	184	229
152 × 76	181	86	197	267
152 × 102	198	106	210	305
152 × 127	217	125	222	343

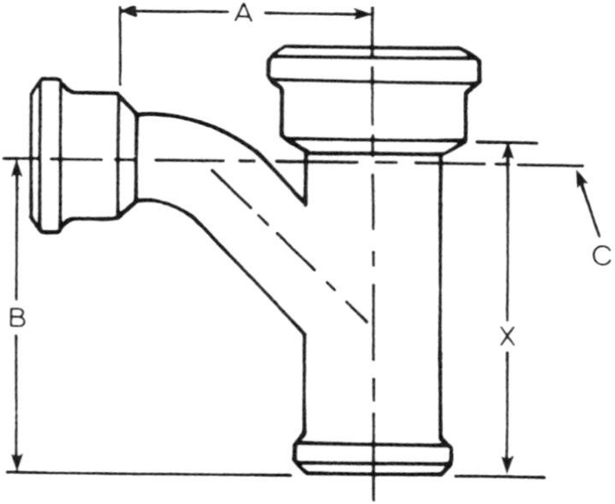

Fig. 14-7A. Single and double combination Y and 1/4 bend.

Table 14-7A. Single and Double Combination Y and 1/4 Bend (Inches)

Size	A	B	X
2	4 7/8	7 3/8	8
3	7	10 1/16	10 1/2
4	9	12	12
5	11	14 1/8	13 1/2
6	12 7/8	16 1/16	15
8	17	21 9/16	19 1/2
3 × 2	5 3/4	8 3/16	9
4 × 2	6 1/4	8 3/16	9
4 × 3	7 1/2	10 1/16	10 1/2
5 × 2	6 3/4	8 3/8	9
5 × 3	8	10 1/16	10 1/2
5 × 4	9 1/2	12	12
6 × 2	7 1/4	8 3/16	9
6 × 3	8 1/2	10 1/16	10 1/2
6 × 4	10	12	12
6 × 5	11 1/2	14 3/16	13 1/2

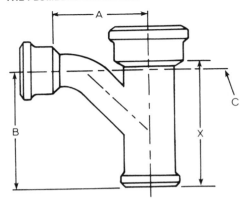

Fig. 14-7B. Single and double combination Y and ¼ bend.

Table 14-7B. Single and Double Combination Y and ¼ Bend (mm)

Size	A	B	X
51	124	187	203
76	178	256	267
102	229	305	305
127	279	359	343
152	327	408	381
203	432	548	495
76 × 51	146	208	229
102 × 51	159	208	229
102 × 76	191	256	267
127 × 51	171	213	229
127 × 76	203	256	267
127 × 102	241	305	305
152 × 51	184	208	229
152 × 76	216	256	267
152 × 102	254	305	305
152 × 127	292	360	343

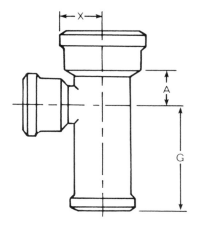

Fig. 14-8A. Single and double T-branches.

Table 14-8A. Single and Double T-Branches (Inches)

Size	X	G	A
2	1¾	6¼	1¾
3	2½	7½	2½
4	3	8	3
5	3½	8½	3½
6	4	9	4
3 × 2	2½	7	2
4 × 2	3	7½	2
4 × 3	3	7	2½
5 × 2	3½	7	2
5 × 3	3½	7½	2½
5 × 4	3½	8	3
6 × 2	4	7	2
6 × 3	4	7½	2½
6 × 4	4	8	3
6 × 5	4	8½	3½

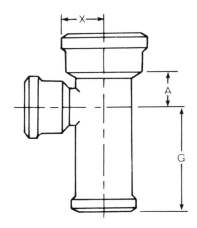

Fig. 14-8B. Single and double T-branches.

Table 14-8B. Single and Double T-Branches (mm)

Size	X	G	A
51	44	159	44
76	64	191	64
102	76	203	76
127	89	216	89
152	102	229	102
76 × 51	64	178	51
102 × 51	76	178	51
102 × 76	76	191	64
127 × 51	89	178	51
127 × 76	89	191	64
127 × 102	89	203	76
152 × 51	102	178	51
152 × 76	102	191	64
152 × 102	102	203	76
152 × 127	102	216	89

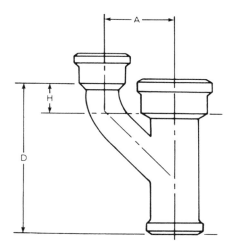

Fig. 14-9. Upright Y branches.

Table 14-9. Upright Y Branches

	Size	A	H	X	D
I	2	4½	2	8	10
n	3	5½	1¹⁵/₁₆	10½	12⁷/₁₆
c	4	6½	1¹⁵/₁₆	12	13¹⁵/₁₆
h	3 × 2	5	1¹⁵/₁₆	9	10¹⁵/₁₆
e	4 × 2	5½	1¹⁵/₁₆	9	10¹⁵/₁₆
s	4 × 3	6	1¹⁵/₁₆	10½	12⁷/₁₆
	51	114	51	203	254
	76	140	49	267	316
m	102	165	49	305	354
m	76 × 51	127	49	229	278
	102 × 51	140	49	229	278
	102 × 76	152	49	267	316

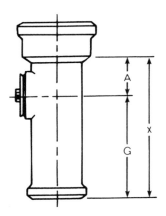

Fig. 14-10. Cleanout T-branch.

Table 14-10. Cleanout T-Branch

Size		A		G		X	
Inches	mm	Inches	mm	Inches	mm	Inches	mm
2	51	1¾	44	6¼	159	8	203
3	76	2½	64	7½	191	10	254
4	102	3	76	8	203	11	279
5	127	3½	89	8½	216	12	305
6	152	4	102	9	229	13	330

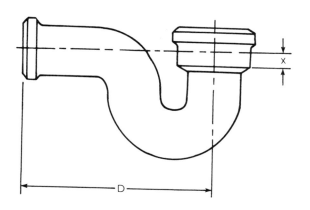

Fig. 14-11. ½ S or P traps.

Table 14-11. ½ S or P Traps

Size		D		X	
Inches	mm	Inches	mm	Inches	mm
2	51	9½	241	1½	38
3	76	12	305	1¼	32
4	102	14	356	1	25
5	127	15½	394	½	13
6	152˙	17	432	0	0

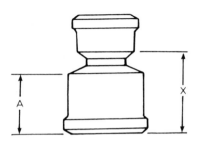

Fig. 14-12. Reducers.

Table 14-12. Reducers

Size		A		X	
Inches	mm	Inches	mm	Inches	mm
3 × 2	76 × 51	3¼	83	4¾	121
4 × 2	102 × 51	4	102	5	127
4 × 3	102 × 76	4	102	5	127
5 × 2	127 × 51	4	102	5	127
5 × 3	127 × 76	4	102	5	127
5 × 4	127 × 102	4	102	5	127
6 × 2	152 × 51	4	102	5	127
6 × 3	152 × 76	4	102	5	127
6 × 4	152 × 102	4	102	5	127
6 × 5	152 × 127	4	102	5	127

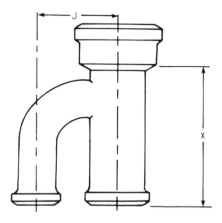

Fig. 14-13. Vent-branches.

Table 14-13. Vent-branches

Size		J		X	
Inches	mm	Inches	mm	Inches	mm
2	51	4½	114	8	203
3	76	5½	140	10	254
4	102	6½	165	11	279
3 × 2	76 × 51	5	127	9	229
4 × 2	102 × 51	5½	140	9	229
4 × 3	102 × 76	6	152	10	254

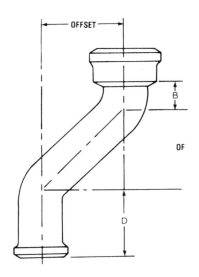

Fig. 14-14A. ⅛ bend offset.

Table 14-14A. ⅛ Bend Offset (Inches)

Size	Offset	Hub	D	B
2	(2-4-6-8-10)	2½	4¼	1
3	(2-4-6-8-10)	2¾	5	1½
4	(2-4-6-8-10)	3	5¼	1¾
5	(2-4-6-8-10)	3	5⁹⁄₁₆	1¹⁵⁄₁₆
6	(2)	3	5⅝	2
6	(4-6-8-10)	3	5¹³⁄₁₆	2³⁄₁₆

Note: ½ bend offset fittings made in pipe sizes 2″ through 6″ (Offsets 2″ through 18″).

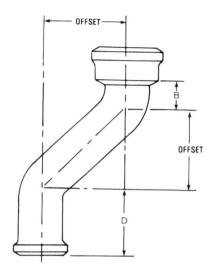

Fig. 14-14B. ⅛ bend offset.

Table 14-14B. ⅛ Bend Offset (mm)

Size	Offset	Hub	D	B
51	(51-102-152-203-254)	64	108	25
76	(51-102-152-203-254)	70	127	38
102	(51-102-152-203-254)	76	133	44
127	(51-102-152-203-254)	76	141	49
152	(51)	76	143	51
152	(102-152-203-254)	76	149	56

Note: ⅛ bend offset fittings made in pipe sizes 51 through 152 mm (Offsets 51 through 457 mm).

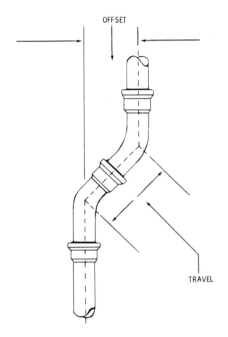

Fig. 14-15. Minimum offsets using ⅙, ⅛, 1/16 bend fittings.

Table 14-15A. Minimum Offsets Using ⅛ Bend Fittings

Size		Travel		Minimum Offset	
Inches	mm	Inches	mm	Inches	mm
2	51	5¾	146	4³⁄₃₂	104
3	76	6⅞	175	4⅞	124
4	102	7⅞	200	5⁹⁄₁₆	141
5	127	8¼	210	5⅞	149
6	152	8⅝	219	6⅛	156
8	203	11⅛	283	7⅞	200
10	254	11⅞	302	8⁷⁄₁₆	214
12	305	14⅜	365	10¼	260

Table 14-15B. Minimum Offsets Using ¹⁄₁₆ Bend Fittings

Size		Travel		Minimum Offset	
Inches	mm	Inches	mm	Inches	mm
2	51	4½	114	1¾	44
3	76	5⅜	137	2¹⁄₁₆	52
4	102	6⅛	156	2⁵⁄₁₆	59
5	127	6¼	159	2⅜	60
6	152	6½	165	2⁷⁄₁₆	62
8	203	8½	216	3¼	83
10	254	8⅞	225	3⅜	86
12	305	11	279	4¼	108

Table 14-15C. Minimum Offsets Using ⅙ Bend Fittings

Size		Travel		Minimum Offset	
Inches	mm	Inches	mm	Inches	mm
2	51	6¾	171	5⅞	149
3	76	8	203	6¹⁵⁄₁₆	176
4	102	9⅛	232	7¹⁵⁄₁₆	202
5	127	9¾	248	8½	216
6	152	10¼	260	8⅞	225
8	203	13⅛	333	11⅜	289
10	254	14¼	362	12⅜	314
12	305	17	432	14¾	375

Cast Iron No-Hub Pipe Fittings and Specifications

The most outstanding advantages of cast iron (€) no-hub joints are: Faster installation, more economical, space-saving [3″ (76 mm) size fits neatly in 2 × 4″ (5 × 10 cm) framed wall], waste: absolutely none; quieter, vibrationless, light-weight, physically less taxing than conventional jointing, testing takes less time. Five floors can be tested at one time.

Introduction

These suggestions are for use with the € no-hub system utilizing a neoprene sleeve-type coupling device consisting of an internally ribbed elastomeric sealing sleeve within a protective corrugated stainless steel shield band secured by two stainless steel bands with tightening devices also of stainless steel.

During installation assembly € no-hub pipe and fittings must be inserted into the sleeve and firmly seated against the center rib or shoulder of the gasket. In order to provide a sound joint with field-cut lengths of pipe it is necessary to have ends cut smooth and as square as possible. Snap- or abrasive-type cutters may be used.

The stainless steel bands must be tightened alternately and firmly to not less than 48, nor more than 60 inch lbs. (67.79 Nm) of torque.

Installation

Vertical Piping

Secure vertical piping at sufficiently close intervals to keep the pipe in alignment and to support the weight of the pipe and its contents. Support stacks at their bases and at each floor.

Horizontal Piping Suspended

Support ordinary horizontal piping and fittings at sufficiently close intervals to maintain alignment and prevent sagging or grade reversal.

Support each length of pipe by a hanger located as near the coupling as possible but not more than 18″ (46 cm) from the joint.

If piping is supported by nonrigid hangers more than 18″ (46 cm) long, install sufficient sway bracing to prevent lateral movement, such as might be caused by seismic shock.

Hangers should also be provided at each horizontal branch connection.

Horizontal Piping Underground

℄ no-hub systems laid in trenches should be continuously supported on undisturbed earth, or compacted fill of selected material, or on masonry blocks at each coupling.

To maintain proper alignment during back filling, stabilize the pipe in the proper position by partial back filling and cradling, or by the use of adequate metal stakes or braces fastened to the pipe.

Piping laid on grade should be adequately staked to prevent misalignment when the slab is poured.

Vertical sections and their connection branches should be adequately staked and fastened to driven pipe or reinforcing rod so as to remain stable while back fill is placed, or concrete is poured.

Remember: The spacer inside of the neoprene gasket where fittings or pipe ends meet measures ³/₃₂ inch (2.38 mm).

Notes: Fitting measurements, laying lengths, may vary ⅛″ (3.2 mm) plus or minus. Five-foot (152 cm) lengths of pipe may vary ¼″ (6.4 mm) plus or minus; 10′ (305 cm) lengths of pipe may vary ½″ (13 mm) plus or minus. All 2″, 3″, and 4″ (51, 76, and 102 mm) fittings, etc., lay the same length; this is an advantage in that one fitting can be removed from a line and another inserted without cutting the pipe. No-hub pipe may be cast with or without a spigot bead and positioning lugs.

Fig. 14-16. No-hub cast iron pipe ready for joining. Note extreme simplicity of joint parts.

Fig. 14-17. Sleeve coupling is placed on one end of pipe. Stainless steel shield and band clamps are placed on the end of the other pipe. Two bands are used on pipe sizes 1½″ to 4″; four bands are used on sizes 5″ through 10″.

Fig. 14-18. Pipe ends are butted against integrally molded shoulder inside of the sleeve. Shield is slid into position and tightened to make a joint that is quickly assembled and permanently fastened.

Fig. 14-19. Fittings are jointed and fastened in the same way. Making up joints in close quarters is easily done.

Fig. 14-20. Extra support is needed with no-hub. Note conventional hangers and supports in the ceiling to make the system more rigid.

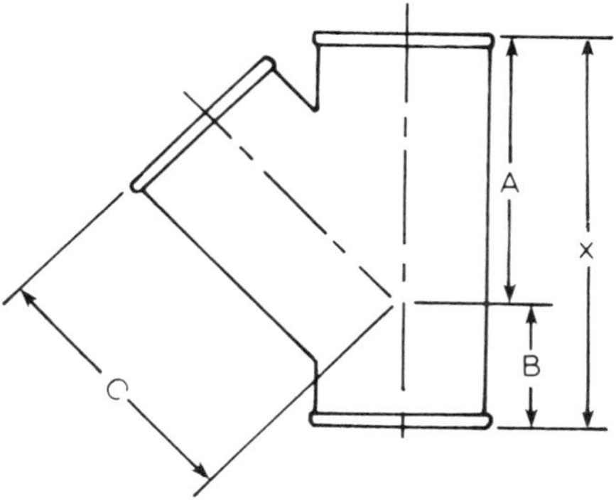

Fig. 14-21A. Y-branches, single and double.

Table 14-16A. Y-Branches, Single and Double (Inches)

Size	A	B	C	X
1½	4	2	4	6
2	4⅝	2	4⅝	6⅝
3	5¾	2¼	5¾	8
4	7 1/16	2 7/16	7 1/16	9½
5	9½	3⅛	9½	12⅝
6	10¾	3 5/16	10¾	14 1/16
3 × 2	5⅛	1½	5 5/16	6⅝
4 × 2	5⅝	1	6	6⅝
4 × 3	6 5/16	1 11/16	6½	8
5 × 2	7⅛	15/16	7½	8 1/16
5 × 3	8	1 11/16	8	9 11/16
5 × 4	8¾	2 7/16	8⅛	11 3/16
6 × 2	7 13/16	½	8¼	8 5/16
6 × 3	8½	1¼	8¾	9¾
6 × 4	9¼	1 15/16	9¼	11 3/16
6 × 5	9 15/16	2 9/16	10¼	12½

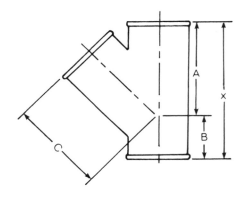

Fig. 14-21B. Y-branches, single and double.

Table 14-16B. Y-Branches, Single and Double (mm)

Size	A	B	C	X
38	102	51	102	152
51	117	51	117	168
76	146	57	146	203
102	179	62	179	241
127	241	79	241	321
152	273	84	273	357
76 × 51	130	38	135	168
102 × 51	143	25	152	168
102 × 76	160	43	165	203
127 × 51	181	24	191	205
127 × 76	203	43	203	246
127 × 102	222	62	216	284
152 × 51	198	13	210	211
152 × 76	216	32	222	248
152 × 102	235	49	235	284
152 × 127	252	64	260	318

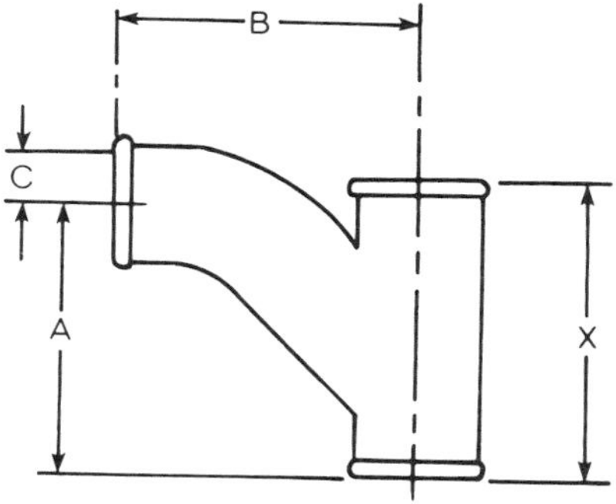

Fig. 14-22A. Single and double combination Y and ¼ bend.

Table 14-17A. Single and Double Combination Y and ¼ Bend (Inches)

Size	A	B	C	X
1½	4¾	5⅜	1¼	6
2	5⅜	6⅛	1¼	6⅝
3	7 5/16	8	11/16	8
4	9¼	10	¼	9½
5	11¾	12½	⅞	12⅝
6	13⅝	14⅜	7/16	14 1/16
2 × 1½	5	5⅝	1	6
3 × 2	5½	6¾	1⅛	6⅝
4 × 2	5½	7¼	1⅛	6⅝
4 × 3	7¼	8½	¾	8
5 × 2	5 15/16	7¾	2⅛	8 1/16
5 × 3	7¾	9	1 15/16	9 11/16
5 × 4	9¾	10½	1 7/16	11 3/16
6 × 2	6	8¼	2 5/16	8 5/16
6 × 3	7 13/16	9½	1 15/16	9¾
6 × 4	9¾	11	1 7/16	11 3/16
6 × 5	11 11/16	13	13/16	12½

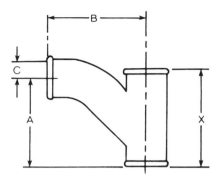

Fig. 14-22B. Single and double combination Y and ¼ bend.

Table 14-17B. Single and Double Combination Y and ¼ Bend (mm)

Size	A	B	C	X
38	121	137	32	152
51	137	156	32	168
76	186	203	17	203
102	235	254	6	241
127	298	318	22	321
152	346	365	11	357
51 × 38	127	143	25	152
76 × 51	140	171	29	168
102 × 51	140	184	29	168
102 × 76	184	216	19	203
127 × 51	151	197	54	205
127 × 76	197	229	49	246
127 × 102	248	267	37	284
152 × 51	152	210	59	211
152 × 76	198	241	49	248
152 × 102	248	279	37	284
152 × 127	297	330	21	318

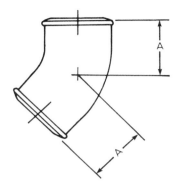

Fig. 14-23. Bends — ⅛, ⅟₁₆, ⅙, ⅕.

Table 14-18A. ⅛ Bends

Size		A	
Inches	mm	Inches	mm
1½	38	2⅝	67
2	51	2¾	70
3	76	3	76
4	102	3⅛	79
5	127	3⅞	98
6	152	4¹⁄₁₆	103

Table 14-18B. ⅟₁₆ Bends

Size		A	
Inches	mm	Inches	mm
2	51	2⅛	54
3	76	2¼	57
4	102	2⁵⁄₁₆	59
5	127	2¹⁵⁄₁₆	75
6	152	3	76

Table 14-18C. ⅙ Bends

Size		A	
Inches	mm	Inches	mm
2	51	3¼	83
3	76	3½	89
4	102	3¹³⁄₁₆	97

Table 14-18D. ⅕ Bends

Size		A	
Inches	mm	Inches	mm
2	51	3¹¹⁄₁₆	94
3	76	4¹⁄₁₆	103
4	102	4⁷⁄₁₆	113

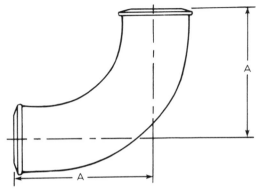

Fig. 14-24. Sweep.

Table 14-19A. Short Sweeps

Size		A	
Inches	mm	Inches	mm
2	51	6½	165
3	76	7	178
4	102	7½	191
5	127	8½	216
6	152	9	229

Table 14-19B. Long Sweeps

Size		A	
Inches	mm	Inches	mm
1½	38	9¼	235
2	51	9½	241
3	76	10	254
4	102	10½	267
5	127	11½	292
6	152	12	305
8	203	13½	343

Table 14-19C. Reducing Long Sweeps

Size		A	
Inches	mm	Inches	mm
3 × 2	76 × 51	10	254
4 × 3	102 × 76	10½	267

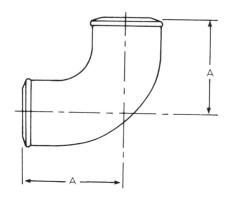

Fig. 14-25. ¼ bend.

Table 14-20. ¼ Bend

Size		A	
Inches	mm	Inches	mm
1½	38	4¼	108
2	51	4½	114
3	76	5	127
4	102	5½	140
5	127	6½	165
6	152	7	178

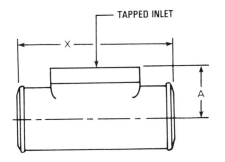

Fig. 14-26. Test tees.

Table 14-21. Test Tees

Size		X		A	
Inches	mm	Inches	mm	Inches	mm
2	51	6⅜	162	2	51
3	76	7¾	197	2¹¹⁄₁₆	68
4	102	8⅞	225	3	76
5	127	11½	292	4½	114
6	152	23½	318	5	127

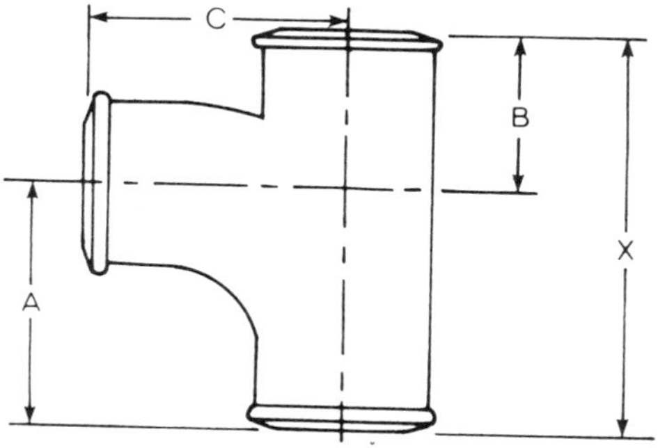

Fig. 14-27A. Single and double sanitary T-branches.

Table 14-22A. Single and Double Sanitary T-Branches (Inches)

Size	A	B	C	X
1½	4¼	2¼	4¼	6½
2	4½	2⅜	4½	6⅞
3	5	3	5	8
4	5½	3⅝	5½	9⅛
3 × 1½	4¼	2¼	5	6½
3 × 2	4½	2⅜	5	6⅞
3 × 4	5½	3½	5	9
4 × 2	4½	2⅜	5½	6⅞
4 × 3	5	3	5½	8
5 × 2	5	3½	6½	8½

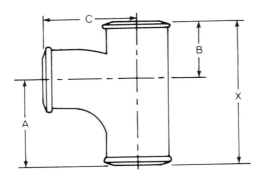

Fig. 14-27B. Single and double sanitary T-branches.

Table 14-22B. Single and Double Sanitary T-Branches (mm)

Size	A	B	C	X
38	108	57	108	165
51	114	60	114	175
76	127	76	127	203
102	140	92	140	232
76 × 38	108	57	127	165
76 × 51	114	60	127	175
76 × 102	140	89	127	229
102 × 51	114	60	140	175
102 × 76	127	76	140	203
127 × 51	127	89	165	216

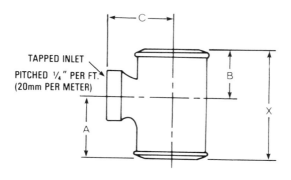

Fig. 14-28A. Single and double sanitary T-branches, tapped.

Table 14-23A. Single and Double Sanitary T-Branches, Tapped (Inches)

Size	A	B	C	X
1½ × 1¼	3¼	2⁷/₁₆	2⁹/₁₆	5¹¹/₁₆
1½ × 1½	3¼	2⁷/₁₆	2⁹/₁₆	5¹¹/₁₆
2 × 1¼	3¼	2⁷/₁₆	2¹³/₁₆	5¹¹/₁₆
2 × 1½	3¼	2⁷/₁₆	2¹³/₁₆	5¹¹/₁₆
2 × 2	3¾	2⁵/₈	3¹/₁₆	6³/₈
3 × 1¼	3¼	2⁷/₁₆	3⁵/₁₆	5¹¹/₁₆
3 × 1½	3¼	2⁷/₁₆	3⁵/₁₆	5¹¹/₁₆
3 × 2	3¾	2⁵/₈	3⁹/₁₆	6³/₈
4 × 1¼	3¼	2⁷/₁₆	3¹³/₁₆	5¹¹/₁₆
4 × 1½	3¼	2⁷/₁₆	3¹³/₁₆	5¹¹/₁₆
4 × 2	3¾	2⁵/₈	4¹/₁₆	6³/₈

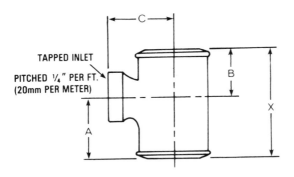

Fig. 14-28B. Single and double sanitary T-branches, tapped.

Table 14-23B. Single and Double Sanitary T-Branches, Tapped (mm)

Size	A	B	C	X
38 × 32	83	62	65	144
38 × 38	83	62	65	144
51 × 32	83	62	71	144
51 × 38	83	62	71	144
51 × 51	95	67	78	162
76 × 32	83	62	84	144
76 × 38	83	62	84	144
76 × 51	95	67	90	162
102 × 32	83	62	97	144
102 × 38	83	62	97	144
102 × 51	95	67	103	162

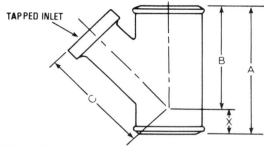

Fig. 14-29. Y-branches, tapped.

Table 14-24A. Y-Branches, Tapped (Inches)

Size	A	B	C	X
2 × 1¼	6⅝	5	5¹/₁₆	1⅝
2 × 1½	6⅝	5	5¹/₁₆	1⅝
2 × 2	6⅝	4⅝	5¹/₁₆	2
3 × 1¼	6⅝	5⅛	5¹/₁₆	1½
3 × 1½	6⅝	5½	5¾	1⅛
3 × 2	6⅝	5⅛	5¹³/₁₆	1½
4 × 1¼	6⅝	5⅝	6⁷/₁₆	1
4 × 1½	6⅝	5⅝	6⁷/₁₆	1
4 × 2	6⅝	5⅝	6½	1

Table 14-24B. Y-Branches, Tapped (mm)

Size	A	B	C	X
51 × 32	168	127	129	41
51 × 38	168	127	129	41
51 × 51	168	117	129	51
76 × 32	168	130	129	38
76 × 38	168	140	146	29
76 × 51	168	130	148	38
102 × 32	168	143	164	25
102 × 38	168	143	164	25
102 × 51	168	143	165	25

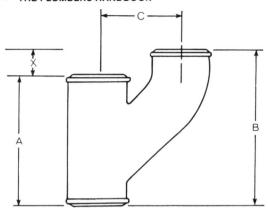

Fig. 14-30. Upright Y.

Table 14-25A. Upright Y (Inches)

Size	A	B	C	X
2	7	10¼	5½	3¼
3	8⅝	10⅝	5½	2¼
4	9¾	11⁹⁄₁₆	6	1¹³⁄₁₆
3 × 2	7	9¹¹⁄₁₆	5½	2¹¹⁄₁₆
4 × 2	7	9¼	5½	2¼
4 × 3	8⅜	10⅛	5½	1¾

Table 14-25B. Upright Y (mm)

Size	A	B	C	X
51	178	260	140	83
76	213	270	140	57
102	248	294	152	46
76 × 51	178	246	140	68
102 × 51	178	235	140	57
102 × 76	213	257	140	44

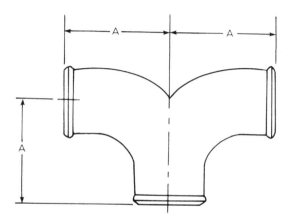

Fig. 14-31. ¼ bend, double.

Table 14-26. ¼ Bend, Double

Size		A	
Inches	mm	Inches	mm
2	51	4½	114
3	76	5	127
4	102	5½	140

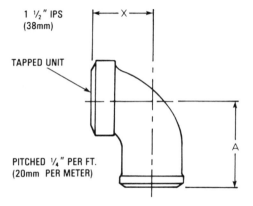

1 ½" IPS (38mm)

TAPPED UNIT

PITCHED ¼" PER FT. (20mm PER METER)

Fig. 14-32. Short radius tapped ¼ bend.

Table 14-27. Short Radius Tapped ¼ Bend

Size		A		X	
Inches	mm	Inches	mm	Inches	mm
1½ × 1¼	38 × 32	3	76	2	51
1½ × 1½	38 × 38	3	76	2	51
2 × 1¼	51 × 32	3¼	83	2¼	57
2 × 1½	51 × 38	3¼	83	2¼	57

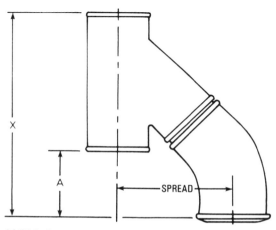

Fig. 14-33A. Spread between vent and revent when using a wye, and ¼ bend, to nearest ¹⁄₁₆.

Table 14-28A. Spread Between Vent and Revent with Wye (Inches)

Size	Spread	X	A
2 × 2	5⁵⁄₁₆	10¹⁄₁₆	3⁷⁄₁₆
3 × 3	6¹⁄₄	11¹⁄₂	3¹⁄₂
4 × 4	7¹⁄₄	12¹²⁄₁₆	3⁵⁄₁₆
3 × 2	5³⁄₄	10	3³⁄₈
4 × 2	6¹⁄₄	10	3³⁄₈
4 × 3	6¹³⁄₁₆	11¹⁄₂	3¹⁄₂
5 × 2	7⁵⁄₁₆	11	2¹⁵⁄₁₆
5 × 3	7⁷⁄₈	12⁹⁄₁₆	2⁷⁄₈
5 × 4	8⁵⁄₁₆	13⁷⁄₈	2¹¹⁄₁₆
6 × 2	7⁷⁄₈	11⁷⁄₈	2¹³⁄₁₆
6 × 3	8³⁄₈	12⁵⁄₈	2⁷⁄₈
6 × 4	8¹³⁄₁₆	13⁷⁄₈	2¹¹⁄₁₆

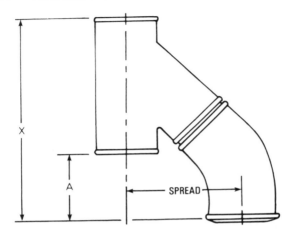

Fig. 14-33B. Spread between vent and revent when using a wye and ¼ bend.

Table 14-28B. Spread Between Vent and Revent with Wye (mm)

Size	Spread	X	A
51 × 51	135	256	87
76 × 76	159	292	89
102 × 102	184	325	84
76 × 51	146	254	86
102 × 51	159	254	86
102 × 76	173	292	89
127 × 51	186	279	75
127 × 76	200	319	73
127 × 102	211	352	68
152 × 51	200	283	71
152 × 76	213	321	73
152 × 102	224	352	68

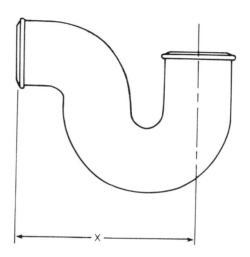

Fig. 14-34. ½ S or P traps.

Table 14-29. ½ S or P Traps

Size		X	
Inches	mm	Inches	mm
1½	38	6¾	171
2	51	7½	191
3	76	9	229
4	102	10½	267
6	152	14	356

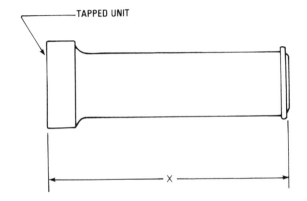

TAPPED UNIT

Fig. 14-35. Tapped extension piece.

Table 14-30. Tapped Extension Piece

Size		X		I.P.S. Tapping	
Inches	mm	Inches	mm	Inches	mm
2	51	12	305	2	51
3	76	12	305	3	76
4	102	12	305	3½	89

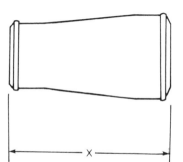

Fig. 14-36. Increaser-reducer.

Table 14-31. Increaser-Reducer

Size		X	
Inches	mm	Inches	mm
1½ × 2	38 × 51	3⅝	92
2 × 3	51 × 76	8	203
2 × 4	51 × 102	8	203
3 × 4	76 × 102	8	203
4 × 6	102 × 152	4	102
5 × 6	127 × 152	4½	114

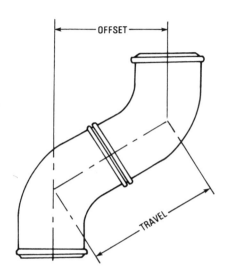

Fig. 14-37. Offset using bends.

Table 14-32A. Minimum Offsets Using No-Hub ⅛ Bend Fittings

Size		Travel		Minimum Offset	
Inches	mm	Inches	mm	Inches	mm
1½	38	5¹¹/₃₂	136	3¹³/₁₆	97
2	51	5¹⁹/₃₂	142	3¹⁵/₁₆	100
3	76	6³/₃₂	155	4⁵/₁₆	110
4	102	6¹¹/₃₂	161	4½	114
5	127	7²⁷/₃₂	199	5⁹/₁₆	141
6	152	8⁷/₃₂	209	5¹³/₁₆	148

8

Table 14-32B. Minimum Offsets Using No-Hub $\frac{1}{16}$ Bend Fittings

Size		Travel		Minimum Offset	
Inches	mm	Inches	mm	Inches	mm
2	51	$4\frac{11}{32}$	110	$1\frac{11}{16}$	43
3	76	$4\frac{19}{32}$	117	$1\frac{3}{4}$	44
4	102	$4\frac{23}{32}$	120	$1\frac{13}{16}$	46
5	127	$5\frac{31}{32}$	152	$2\frac{5}{16}$	59
6	152	$6\frac{3}{32}$	155	$2\frac{5}{16}$	59

Table 14-32C. Minimum Offsets Using No-Hub $\frac{1}{6}$ Bend Fittings

Size		Travel		Minimum Offsets	
Inches	mm	Inches	mm	Inches	mm
2	51	$6\frac{19}{32}$	167	$5\frac{11}{16}$	144
3	76	$7\frac{3}{32}$	180	$6\frac{1}{8}$	156
4	102	$7\frac{23}{32}$	196	$6\frac{11}{16}$	170

Table 14-32D. Minimum Offsets Using No-Hub $\frac{1}{5}$ Bend Fittings

Size		Travel		Minimum Offset	
Inches	mm	Inches	mm	Inches	mm
2	51	$7\frac{15}{32}$	190	$7\frac{1}{16}$	179
3	76	$8\frac{7}{32}$	209	$7\frac{13}{16}$	198
4	102	$8\frac{31}{32}$	228	$8\frac{1}{2}$	216

Note: Minimum inch offsets figured to nearest $\frac{1}{16}$"; minimum metric offsets to nearest millimeter.

Fig. 14-38. Sway brace and method of hanging.

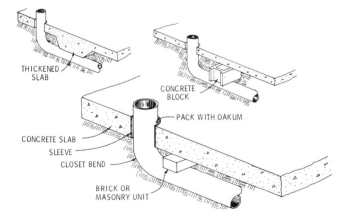

THICKENED SLAB

CONCRETE BLOCK

PACK WITH OAKUM

CONCRETE SLAB

SLEEVE

CLOSET BEND

BRICK OR MASONRY UNIT

Fig. 14-39.

Fig. 14-40. Note sway brace and the method of hanging and cleanout.

Fig. 14-41. Note the neat method of hanging pipe.

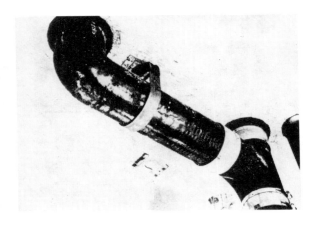

Fig. 14-42. Method of using hanger for a closet bend. Note sleeves and oakum in sleeves.

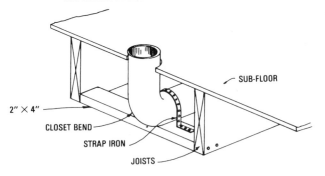

SUB-FLOOR

2" × 4"

CLOSET BEND

STRAP IRON

JOISTS

Fig. 14-43. Bracing for a closet bend.

Fig. 14-44. View showing method of hanging and sway bracing.

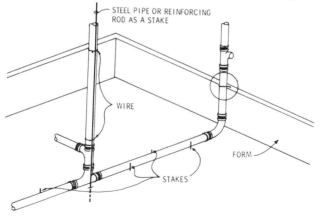

STEEL PIPE OR REINFORCING
ROD AS A STAKE

WIRE

FORM

STAKES

Fig. 14-45. Slab-on-grade installation.

Fig. 14-46. Method of clamping the no-hub pipe at each floor, using a friction clamp or floor clamp.

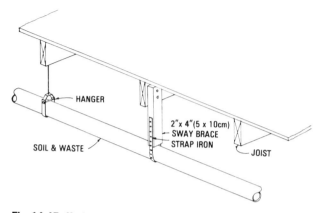

Fig. 14-47. Horizontal pipe with sway brace.

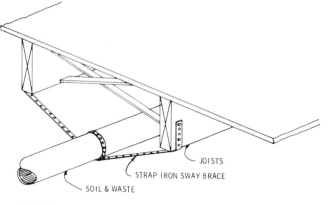

Fig. 14-48. Sway brace.

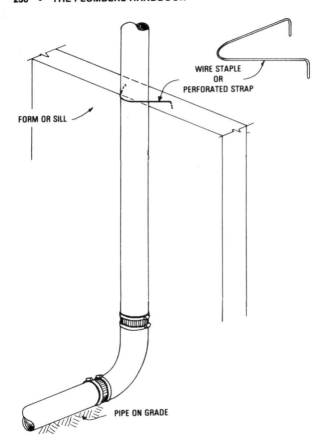

WIRE STAPLE
OR
PERFORATED STRAP

FORM OR SILL

PIPE ON GRADE

Fig. 14-49. Support for vertical pipe.

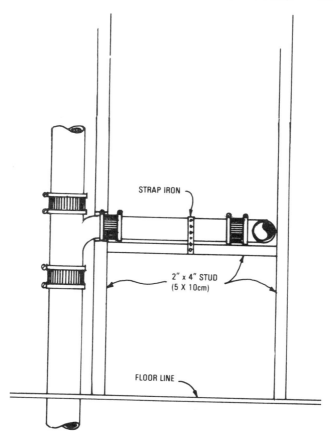

Fig. 14-50. Strapping horizontal run to a cross brace.

15
Copper Pipe and Fittings

Types of Copper Tube

There are two principal types of copper tube: plumbing tube and ARC tube. Plumbing tube includes types K, L, M, and DWV. ARC tube is used for air conditioning and refrigeration. Table 15-1 lists the classes and types available.

Types K, L, M, and DWV come in standard sizes with the outside diameter always ⅛″ (.32 cm) larger than the standard size. Each type represents a series of sizes with different wall thicknesses. Inside diameters depend on tube size and wall thickness. Drawn tube is hard tube, and annealed tube is soft. Hard tubing can be joined by soldering or brazing, using capillary fittings, or by welding. Tube in the bending or soft tempers can be joined in the same ways, as well as with flare-type compression fittings. It is also possible to expand the end of one tube so that it can be joined to another by soldering or brazing without a capillary fitting.

Recommendations for Various Applications

Strength, formability, and other mechanical factors frequently determine the type of copper tube to be used in a particular application. Sometimes building or plumbing codes govern what types may be used. When a choice can be made, it is helpful to know which type of copper tube has served and will serve successfully and economically in the following applications.

Table 15-1. Standard Copper Plumbing Tube

COMMERCIALLY AVAILABLE LENGTHS		
TUBE	DRAWN	ANNEALED
Type K Available in diameters from $\frac{1}{4}$ to 12" or .64 to 30.5 cm	Straight Lengths: Up to 8" (20.3 cm) S.P.S. 20 ft. (6.1 m) 10" (25.4 cm) 18 ft. (5.49 m) 12" (30.5 cm) 12 ft. (3.66 m)	Straight Lengths: Up to 8" (20.3 cm) 20 ft. (6.1 m) 10" (25.4 cm) 18 ft. (5.49 m) 12" (30.5 cm) 12 Ft. (3.66 m) Coils: Up to 1" (2.54 cm) S.P.S. 60 ft. (18.29 m) 100 ft. (30.48 m) $1\frac{1}{4}$ & $1\frac{1}{2}$ (3.18 & 3.8 cm) 60 ft. (18.29 m) 40 ft. (12.19 m) 2" (5.1 cm) 45 ft. (13.72 m)
Type L Available in diameters from $\frac{1}{4}$ to 12" or .64 to 30.5 cm	Straight Lengths: Up to 10" (25.4 cm) S.P.S. 20 ft. (6.1 m) 12" (30.5 cm) 18 ft. (5.49 m)	Straight Lengths: Up to 10" (25.4 cm) 20 ft. (6.1 m) 12" (30.5 cm) 18 ft. (5.49 m) Coils: Up to 1" (2.54 cm) 60 ft. (18.29 m) 100 ft. (30.48 m) $1\frac{1}{4}$ & $1\frac{1}{2}$" (3.18 & 3.8 cm) 60 ft. (18.29 m) 2" (5.1 cm) 40 ft. (12.19 m) 45 ft. (13.72 m)
Type M Available in diameters from $\frac{3}{8}$ to 12" or .95 to 30.5 cm	Straight Lengths: All diameters 20 ft. (6.1 m)	Straight Lengths: Up to 12" (30.5 cm) 20 ft. (6.1 m) Coils: Up to 1" (2.54 cm) 60 ft. (18.29 m) 100 ft. (30.48 m) $1\frac{1}{4}$ & $1\frac{1}{2}$" (3.18 & 3.8 cm) 60 ft. (18.29 m) 2" (5.1 cm) 40 ft. (12.19 m) 45 ft. (13.72 m)

Table 15-1. Standard Copper Plumbing Tube (Continued)

COMMERCIALLY AVAILABLE LENGTHS		
TUBE	DRAWN	ANNEALED
DWV Available in diameters from 1¼″ to 8″ or 3.18 to 20.3 cm	Straight Lengths: All diameters 20 ft. (6.1 m)	Not available
ACR Available in diameters from ⅛ to 4⅛″ or .32 to 10.48 cm	Straight Lengths: 20 ft. (6.1 m)	Coils: 50 ft. (15.2 m)

Underground water services. Use Type M for straight lengths joined with fittings, and Type L soft temper where coils are more convenient.

Water distribution systems. Use Type M for above and below ground.

Chilled water mains. Use Type M for sizes up to 1 inch (2.54 cm) and Type DWV for sizes of 1¼″ (3.18 cm) and larger.

Drainage and vent systems. Use Type DWV for above and below ground, waste, soil, and vent lines, roof drainage, building drains, and building sewers.

Heating. For radiant panel and hydronic heating, and for snow melting systems, use Type L soft temper where coils are formed in place or prefabricated, Type M where straight lengths joined with fittings are used. For hot-water heating and low-pressure steam, use Type M for sizes up to 1¼″ (3.18 cm) and Type DWV for sizes of 1¼″ and larger. For condensate return lines, Type L is successfully used.

Fuel oil and underground gas services. Use copper tube in accord with local codes.

Oxygen systems. Use Type L or K, suitably cleaned for oxygen service per NFPA.*

*National Fire Protection Association, 470 Atlantic Avenue, Boston, MA 02210.

Installation Tips

Small Systems

Distribution systems for single-family houses can be sized easily on the basis of experience and any applicable code requirements, as can other similar small installations. Detailed study of the six design considerations above is not necessary in such cases. The size of the short branches to individual fixtures can be determined by reference to Table 15-2. In general, the mains servicing these fixture branches can then be sized as follows:

Table 15-2. Examples of Minimum Copper Tube Sizes for Short-Branch Connections to Fixtures

Fixture	Copper Tube Size, inches (cm)
Drinking Fountain	⅜ (0.95)
Lavatory	⅜ (0.95)
Water Closet (tank type)	⅜ (0.95)
Bathtub	½ (1.27)
Dishwasher (home)	½ (1.27)
Kitchen Sink (home)	½ (1.27)
Laundry Tray	½ (1.27)
Service Sink	½ (1.27)
Shower Head	½ (1.27)
Sill Cock, Hose Bibb, Wall Hydrant	½ (1.27)
Urinal (tank type)	½ (1.27)
Washing Machine (home)	½ (1.27)
Kitchen Sink (commercial)	¾ (1.91)
Urinal (flush valve)	¾ (1.91)
Water Closet (flush valve)	1 (2.54)

1. Up to three ⅜″ (0.95 cm) branches can be served by a ½″ (1.27 cm) main.
2. Up to three ½″ (1.27 cm) branches or up to five ⅜″ (0.95 cm) branches can be served by a ¾″ (1.91 cm) main.

3. Up to three ¾" (1.91 cm) branches or correspondingly more ½" (1.27 cm) or ⅜" (0.95 cm) branches can be served by a 1" (2.54 cm) main.

Generous sizing within these limits will give good design and the best service. Working to minimum sizing within these guidelines will give an adequate system most of the time, depending on available main pressure and the probability of simultaneous use of fixtures. The water distribution system in many single-family homes with 2½ baths, for example, has been completely plumbed with copper tube of ¾" size (1.91 cm) and smaller. The sizing of more complex distribution systems requires detailed analysis of each of the size design considerations listed above.

Pressure Considerations

The water service pressure at the point where the building distribution system (or segment or zone thereof) begins depends either on the local main pressure, the local code, the pressure desired by the system designer, or on a combination of these. In any case, it should not be higher than about 80 psi (552 kPa).

Water Demand

Each fixture in the system represents a certain demand for water. Some examples of approximate water demand in gallons per minute (gpm), or liters per minute (lpm) of flow, are:

	lpm	gpm
Drinking fountain	2.8	0.75
Lavatory faucet	7.6	2
Lavatory faucet, self-closing	9.5	2.5
Sink faucet, WC tank ball cock	11.4	3
Bathtub faucet, shower head, laundry tub faucet	15.1	4
Sill cock, hose bibb, wall hydrant	18.9	5
Flush valve (depending on design)	57-132	15-35

Adding up numbers like these to cover all the fixtures in an entire building distribution system would give the total demand for water usage, in gallons per minute, if all the fixtures were operating and flowing at the same time, which of course does not happen. A reasonable estimate of demand is one based on the extent to which various fixtures in the building might actually be used simultaneously.

Pressure Losses Due to Friction

The pressure available to move the water through the distribution system (or a part thereof) is the main pressure minus: (1) the pressure loss in the meter, (2) the pressure needed to lift water to the highest fixture (static pressure loss), and (3) the pressure needed at the fixtures themselves. This remaining available pressure must be adequate to overcome the pressure losses due to friction during flow of the total demand (intermittent plus continuous fixtures) through the distribution system and its various parts. The final operation is to select the tube sizes in accordance with the pressure losses due to friction. In actual practice the design operation may involve repeating the steps in the design process to readjust pressure, velocity, and size to achieve the best balance of main pressure, tube size, velocity, and available pressure at the fixtures for the design flow required in the various parts of the system.

Fig. 15-1 shows a collection of various type fittings and adapters. Fig. 15-2 shows photos and drawings of specific drainage fittings.

Copper Sovent

"The single-stack plumbing system," as the Sovent is called, was invented in 1959 by Fritz Sommer of Bern, Switzerland. It is a single-stack plumbing system designed to improve and simplify soil, waste, and vent plumbing in multi-story buildings.

Fig. 15-1. Assorted fittings and adapters. *(Courtesy NIBCO, Inc.)*

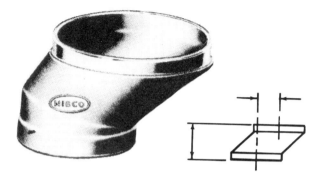

OFFSET CLOSET FITTING—FTGXC

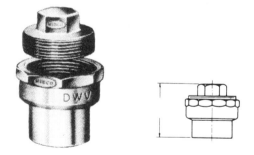

DWV—FTG. X CLEANOUT W/PLUG

Fig. 15-2. Copper drainage fittings.

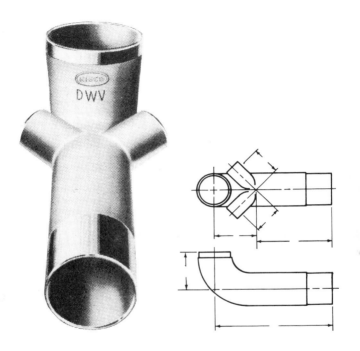

DWV CLOSET BEND — RIGHT & LEFT INLET

Fig. 15-2. Continued.

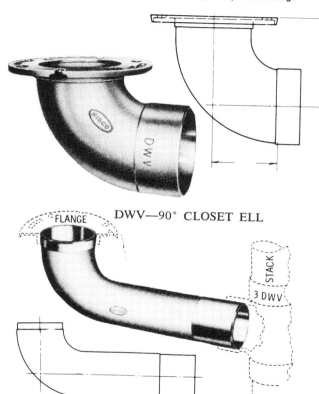

DWV—90° CLOSET ELL

DWV CLOSET BEND FTG. X FTG.

Fig. 15-2. Continued.

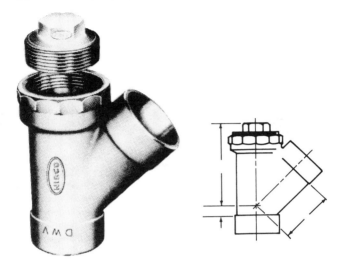

DWV — 45° Y — WITH C.O. PLUG

DWV — 45° — FTG. X COPPER ELL

Fig. 15-2. Continued.

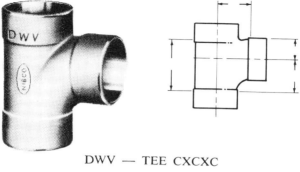

DWV — TEE CXCXC

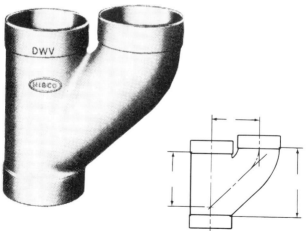

DWV UPRIGHT WYE — CXCXC

Fig. 15-2. Continued.

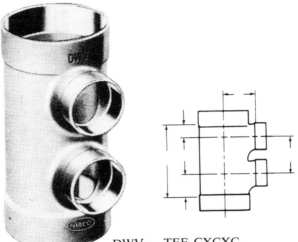

DWV — TEE CXCXC
DWV STACK FITTING — W/2-SIDE INLETS

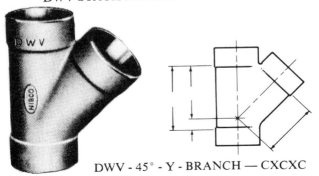

DWV - 45° - Y - BRANCH — CXCXC

Fig. 15-2. Continued.

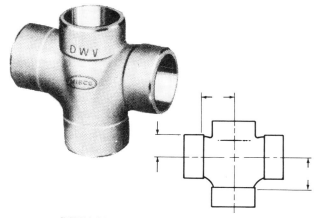

DWV VENT CROSS CXCXCXC

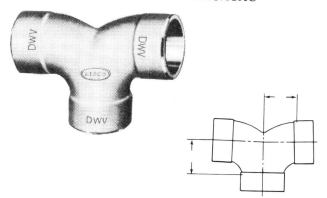

DWV TWIN ELL CXCXC

Fig. 15-2. Continued.

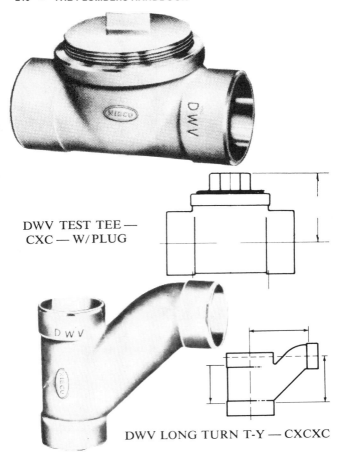

DWV TEST TEE — CXC — W/PLUG

DWV LONG TURN T-Y — CXCXC

Fig. 15-2. Continued.

The basic design rules illustrated here are based on experience gained in the design and construction of hundreds of Sovent systems serving thousands of living units, not to mention the extensive experimental work conducted on the 10-story plumbing test tower.

The first Sovent installation was made in 1961 in Bern, Switzerland. Eight years later 15,000 apartment units were installed in 200 buildings, up to thirty stories in height. Through the Copper Development Association, Inc., extensive tests were carried out on the instrumented test tower. Following the successful completion of these early tests, the system was brought to America.

The first installation was the Habitat Apartments constructed for Expo '67 in Montreal; the next, the Uniment structure in Richmond, California. In 1970, two large apartment buildings and a 10-story office building using Sovent were begun.

Each individual Sovent system must be designed to meet the conditions under which it will operate, and engineering judgment is required in applying the basic design rules presented here to specific buildings. The Copper Development Association, Inc. will be pleased to review Sovent system designs to help ensure that the design principles are followed.

The copper Sovent plumbing system has four parts:

1. A copper DWV stack.
2. A Sovent aerator fitting at each floor level.
3. Copper DWV horizontal branches.
4. A Sovent deaerator fitting at the base of the stack and at the upstream end of each horizontal offset.

The two special fittings, the *aerator* and the *deaerator*, are the basis for the self-venting features of Sovent. It is claimed that a Sovent system will handle at least the same drainage fixture load as a conventional stack of the same diameter, but without the need for the separate vent stack and fixture re-vents needed in the traditional systems.

The aerator does three things:

1. It limits the velocity of both liquid and air in the stack.
2. It prevents the cross section of the stack from filling with a plug of water.
3. It efficiently mixes the waste flowing in the branches with the air in the stack.

The Sovent aerator fitting is said to mix waste and air so effectively that no plug of water can form across the stack diameter and disturb fixture trap seals.

At a floor level where no aerator fitting is needed, a double in-line offset is used, as shown in Fig. 15-3.

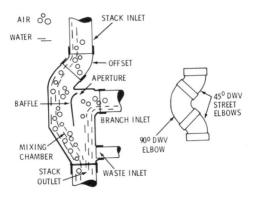

Fig. 15-3. Double in-line offset.

Aerator Fittings

The Sovent aerator consists of an offset at the upper-stack inlet connection, a mixing chamber, one or more branch inlets, one or more waste inlets for connection of smaller waste branches, a baffle in the center of the chamber with an aperture between it and the top of the fitting, and the stack outlet at the bottom of the fitting. Waste branches connect to the side and face inlets on the aerator.

The aerator fitting provides a chamber where the flow of soil and waste from horizontal branches can unite smoothly with the air and liquid already flowing in the stack. The aerator fitting prevents pressure and vacuum fluctuations that could blow or siphon fixture trap seals.

No aerator fitting is needed at a floor level where no soil branch enters and only a 2″ (51 mm) waste branch enters a 4″ (102 mm) stack. A double in-line offset is used in place of the aerator fitting. This offset reduces the fall rate in the stack between floor intervals in a manner similar to the aerator fitting (Fig. 15-4).

The Sovent deaerator fitting relieves the pressure buildup at the bottom of the stack, venting that pressure into a relief line that connects into the top of the building drain. The deaerator pressure relief line is tied to the building drain or at an offset to the lower stack. See Fig. 15-5, and also related drawing Fig. 15-4.

Deaerator Fittings

The Sovent deaerator consists of an air-separation chamber having an internal nosepiece, a stack inlet, a pressure relief outlet at the top, and a stack outlet at the bottom. The deaerator fitting at the bottom of the stack functions in combination with the aerator fitting above to make the single stack self-venting.

The deaerator is designed to overcome the tendency that would otherwise occur for the falling waste to build up excessive back pressure at the bottom of the stack when the flow is decelerated by the bend into the horizontal drain. The deaerator provides a method of separating air from system flow and equalizes pressure buildups.

Tests show that the configuration of the deaerator fitting causes part of the air falling with the liquid and solid in the stack to be ejected through the pressure relief line to the top of the

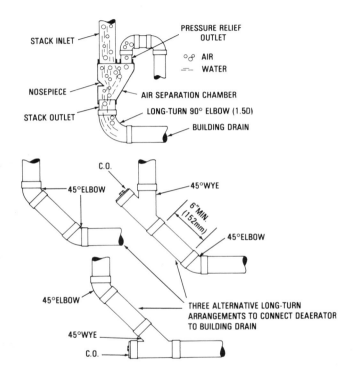

Fig. 15-4. Aerator fitting and "long-turn" details.

building drain, while the balance goes into the drain with the soil and waste.

At the deaerator outlet, the stack is connected into the horizontal drain through a long-turn fitting arrangement. Down-

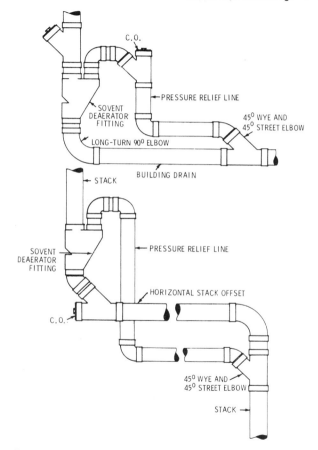

Fig. 15-5. Pressure-relief system.

stream at least 4′ (122 cm) from this point the pressure relief line
from the top of the deaerator fitting is connected into the top of
the building drain.

A deaerator fitting, with its pressure-relief line connection, is
installed in this way at the base of every Sovent stack and also at
every offset (vertical-horizontal-vertical) in a stack. In the latter
case, the pressure-relief line is tied into the stack immediately
below the horizontal portion (Fig. 15-6).

Stacks

The stack must be carried full size through the roof. Two
stacks can be tied together at the top above the highest fixture,
with one stack extending through the roof. If the distance
between the two stacks is 20′ (6.1 m) or less, the horizontal
tie-line can be the same diameter as the stack that terminates
below the roof level. If the distance is greater than 20′ (6.1 m)
the tie-line must be one size larger than the terminated stack.

The common stack extending through the roof must be one
pipe size larger than the size of the larger stack below their
tie-line.

The size of the stack is determined by the number of fixture
units connected, as with traditional sanitary systems. Existing
Sovent installations include 4″ (102 mm) stacks serving up to
400 fixture units and 30 stories in height.

The Sovent's cost-saving potential can be seen by consider-
ing the 12-story stack illustration serving an apartment group-
ing. The material saving is shown graphically in the schematic
riser diagrams for two-pipe and Sovent systems (Fig. 15-7).

Sovent stacks are anchored in the same manner as others.
Noise can be held to a minimum by making sure that no stack or
branch is touching or bearing on a beam, brace, stud, or any
other structural member. Sewer and waste lines can be tied as
shown in Figs. 15-8 and 15-9.

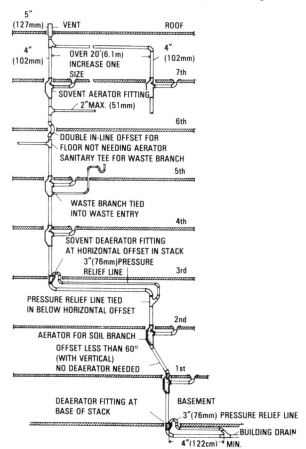

Fig. 15-6. Sovent drainage stack design features.

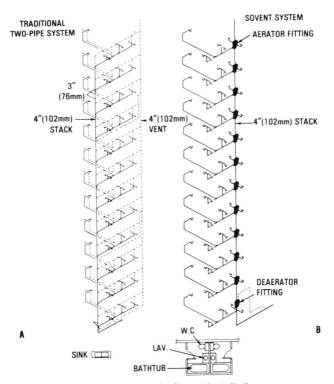

Fig. 15-7. Schematic drawing of a Sovent installation.

Soil and waste branches may be connected into a horizontal stack offset. Waste branches may be connected into the pressure relief line (Figs. 15-10 and 15-11).

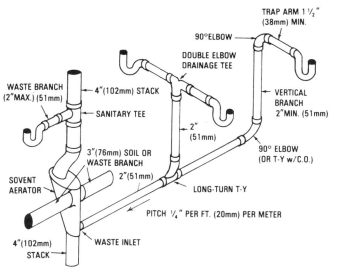

Fig. 15-8. Typical sewer and waste tie-ins.

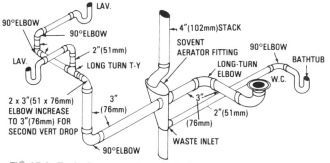

Fig. 15-9. Typical sewer and waste tie-ins.

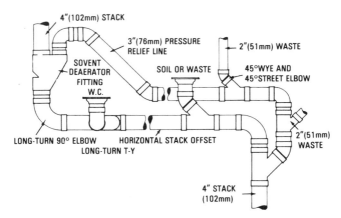

Fig. 15-10. Sewer and waste tie-ins to horizontal stack offset.

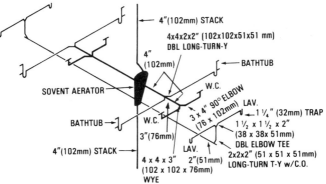

Fig. 15-11. Schematic of sewer and waste connections to a vertical stack.

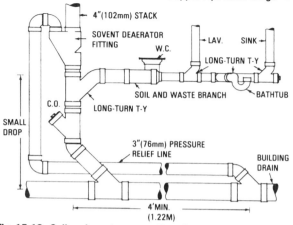

Fig. 15-12. Soil and waste connections just below deaerator at the bottom of the stack.

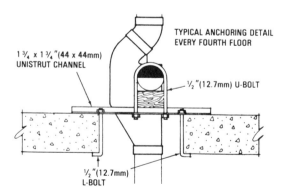

Fig. 15-13. Sovent stack anchoring.

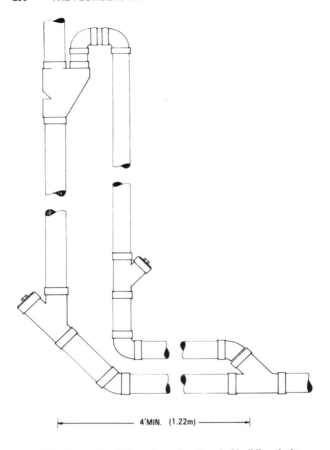

4'MIN. (1.22m)

Fig. 15-14. Deaerator fitting above floor level of building drain.

Fig. 15-15. Installing a copper Sovent system.

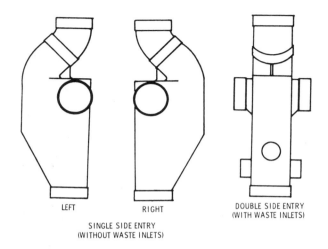

LEFT   RIGHT

SINGLE SIDE ENTRY
(WITHOUT WASTE INLETS)

DOUBLE SIDE ENTRY
(WITH WASTE INLETS)

Fig. 15-16. Sovent fittings with single- and double-side entries.

Soil and waste branches may be connected immediately below a deaerator fitting at the bottom of the stack. The deaerator fitting may be located above the floor level of the building drain.

16
Water Heaters

The DVE and the DRE models of the A.O. Smith Corp. are shown in Fig. 16-1A. They have phase convertibility, i.e., they can be changed over from single-phase to operate on a three-phase electrical power supply and vice versa, in the field, without rewiring.

Installation

A water heater is best located near a floor drain. Flushing and draining the tank is easier when there is a drain nearby. The discharge opening of the relief valve should always be piped to an open drain. In the interest of energy conservation, it is a good idea to keep hot water piping and branch circuit wiring as short as possible, and to insulate hot and cold water piping where heat loss and condensation may be a problem. The heater should be located in an area where leakage of the tank or connections will not result in damage to the area adjacent to the heater or to lower floors of the structure.

When such locations cannot be avoided, a suitable drain pan should be installed under the heater. The pan should be at least 2″ greater than the length and width of the heater, and should be piped to an adequate drain. Suggested clearances from adjacent surfaces are 18″ (46 cm) in front for access to the controls and elements. The heater may be installed on or against combustible surfaces. The temperature of the space in which the heater is installed should not go below 32°F (0°C).

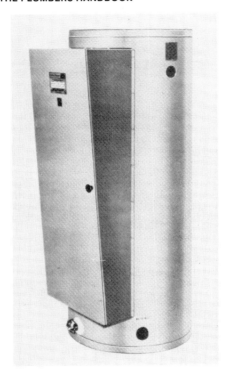

Fig. 16-1A. The DVE and DRE Conservationists are two energy-saving electric water heaters. The manufacturer reports that both can reduce energy costs by up to 43 percent and meet ASHRAE 90-75 Standard for energy efficiencies.
(Courtesy A.O. Smith Corporation, Consumer Products Division)

Fig. 16-1B. The BTP models of water heater are gas-filled. *(Courtesy A.O. Smith Corporation Products Division)*

Water heater corrosion and component failure can be caused by the heating and breakdown of airborne chemical vapors. Spray can propellants, cleaning solvents, refrigerator and air conditioning refrigerants, swimming pool chemicals, calcium and sodium chloride, waxes, and process chemicals are typical compounds that are potentially corrosive. These materials are corrosive at very low concentration levels with little or no odor to reveal their presence. Products of this sort should not be stored near the heater. Also, air which is brought in contact with the water heater should not contain any of these chemicals.

The heater water inlet is usually located on the side of the heater near the bottom. The heater outlet is located at the top of the heater. Fig. 16-2 is a piping diagram.

An unplugged 3/4″ (1.91 cm) relief valve opening is provided for installing a listed temperature and pressure relief valve.

Install temperature and pressure protective equipment required by local codes. The pressure setting of the relief valve should not exceed the pressure capacity of any component in the system. The temperature setting of the relief valve should not exceed 210°F (98.9°C).

Gas-fired water heaters may be installed on a combustible floor; however, clearance to adjacent surfaces must be provided as shown in Fig. 16-3. Units that are to be installed on combustible flooring must be supported by a full layer of hollow concrete blocks from 8 to 12″ (20.32 to 30.48 cm) thick and extending 12″ (30.48 cm) (minimum) beyond the heater in all directions. The concrete blocks must provide an unbroken concrete surface under the heater, with the hollows running continuously and horizontally. If electrical conduits run around the floor of the proposed heater location, insulate the floor as recommended above.

Some models are shipped with the burner installed on the heater. If the burner is shipped separately from the heater, it should be installed after the handling and leveling of the unit has been accomplished.

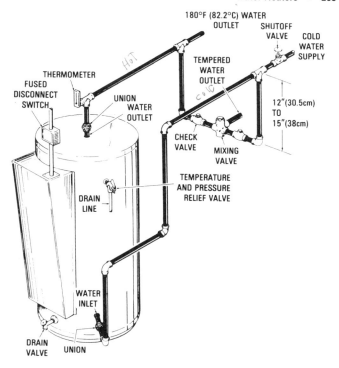

180°F (82.2°C) WATER OUTLET

SHUTOFF VALVE

COLD WATER SUPPLY

TEMPERED WATER OUTLET

THERMOMETER

HOT

FUSED DISCONNECT SWITCH

UNION WATER OUTLET

COLD

12″(30.5cm) TO 15″(38cm)

CHECK VALVE

MIXING VALVE

TEMPERATURE AND PRESSURE RELIEF VALVE

DRAIN LINE

WATER INLET

DRAIN VALVE

UNION

Fig. 16-2A. Water heater—two temperature with mixing valve. (Courtesy A.O. Smith Corporation, Consumer Products Division)

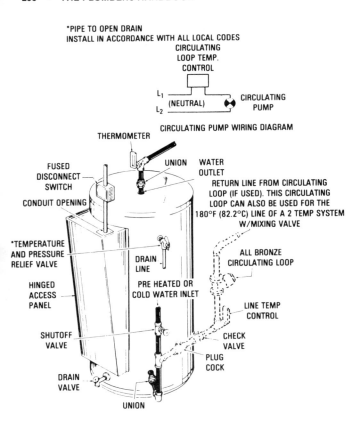

Fig. 16-2B. Water heater—one temperature. *(Courtesy A.O. Smith Corporation, Consumer Products Division)*

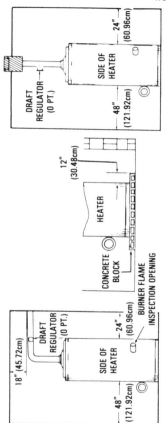

Fig. 16-3. Gas-fired heater installation. (Courtesy A.O. Smith Corporation, Consumer Products Division)

Assemble the burner and gaskets into the tank as shown in Fig. 16-4. The gas control string assembly may be shipped in the vertical position. When installing the unit, the gas control string must be in the horizontal position.

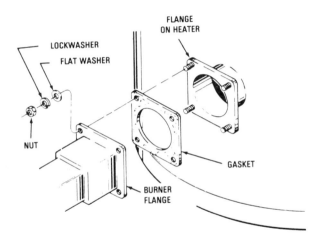

Fig. 16-4. Burner installation. *(Courtesy A.O. Smith Corporation, Consumer Products Division)*

The water heater area should have sufficient air for satisfactory combustion of gas, proper venting, and maintenance of safe ambient temperature. The power burner maker's instructions should be followed with regard to the size of combustion and ventilation air openings. The opening size is based on the vent connector being equipped with a draft regulator. When the heater is installed in an area in which exhaust or ventilating fans may create unsatisfactory combustion or venting, approved provisions must be made to overcome the problem.

The chimney vent connector diameter should be the same size as the heater flue outlet (Table 16-1). A minimum rise of ¼″ (0.64 cm) per foot (30.48 cm) of horizontal connector length must be maintained between the heater and chimney opening, as shown in Fig. 16-5. The connector length should be kept as short as possible.

Table 16-1. Vent Connector Diameter

BTP Model Number	Flue Outlet Inches (Centimeters)
200-300	6 (15.24)
200-600 & 400-600	8 (20.32)
200-800, 200-1000, 400-800 & 400-1000	10 (25.4)
200-1250, 200-1500, 400-1250 & 400-1500	12 (30.48)
400-1750 & 400-2000	14 (35.56)
400-2250 & 400-2500	16 (40.64)

Connectors must not be connected to a chimney vent or venting system served by a power exhauster unless the connection is made on the negative-pressure side of the exhauster. A draft regulator may be installed in the same room as the heater (Fig. 16-5). Locate the regulator as close as possible to the heater and at least 18″ (45.72 cm) from a combustible ceiling or wall. A manually operated damper should not be placed in the chimney connector.

Figs. 16-6 through 16-9 suggest typical methods of making water line connections to the heater. When a circulating pump is used, it is important to have a plug cock in the piping after the pump to regulate the water flow, preventing turbulence in the tank.

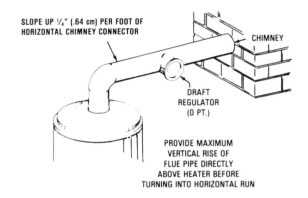

SLOPE UP ¼" (.64 cm) PER FOOT OF
HORIZONTAL CHIMNEY CONNECTOR

CHIMNEY

DRAFT
REGULATOR
(0 PT.)

PROVIDE MAXIMUM
VERTICAL RISE OF
FLUE PIPE DIRECTLY
ABOVE HEATER BEFORE
TURNING INTO HORIZONTAL RUN

Fig. 16-5. Vent connection to a chimney. *(Courtesy A. O. Smith Corporation, Consumer Products Division)*

The power burner maker's instructions should be followed with regard to the size and arrangement of the gas piping. The gas piping schematic shows field- and factory-installed piping and component locations at the power burner.

All electrical work must be installed in accordance with the *National Electrical Code®* and local requirements. An electrical ground is required to reduce the risk of electrical shock. Do not energize the branch circuit before the heater tank is filled with water. A separate gas burner wiring diagram is supplied with the heater literature.

To light the gas burner:

1. Turn the gas cock handle on the control to the off position and dial the assembly to the lowest temperature position.

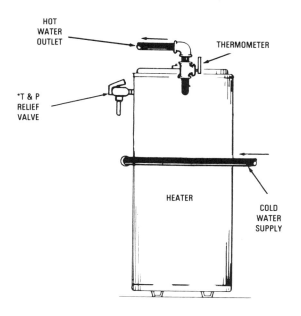

Fig. 16-6. Water line connection to a single-temperature model.
(Courtesy A.O. Smith Corporation, Consumer Products Division)

2. Wait approximately 5 minutes to allow gas that may have accumulated in the burner compartment to escape.
3. Turn the gas cock handle on the control to the pilot position.
4. Fully depress the set button and light the pilot burner.

5. Allow the pilot to burn approximately 1 minute before releasing the set button. If the pilot does not remain lighted, repeat the operation.
6. Turn the gas cock handle on the control to the on position and turn the dial assembly to the desired position. The main burner will then ignite. Adjust the pilot burner air shutter (if provided) to obtain a soft blue flame.

Solar System Water Heaters

America's dwindling supply of energy is becoming more critical as each new year passes. And with the cost of electricity and gas rising, it may be wise for all of us to take a closer look at solar energy, or at least become familiar with a system that is probably here to stay.

The Conservationist™ Solar System Water Heater — illustrated on the following pages — can be tailored to most areas and prevailing conditions. A. O. Smith solar systems can be designed for use with existing hot water heaters.

The Conservationist Solar System Water Heater

Here's how the A. O. Smith Conservationist solar system works:

The hot sun rays are absorbed by roof-mounted collector panels to heat special antifreeze fluid that is circulating through integral copper channels.

A closed-loop system is used for transfer of the heated solution and return. Propylene glycol eliminates any worries of freezing.

A heater-mounted differential controller has a modulating output to collect the maximum amount of available heat from collector panels, even on cloudy days. The pump is adjustable for flow with a restrictor that makes this solar system flexible for various installations.

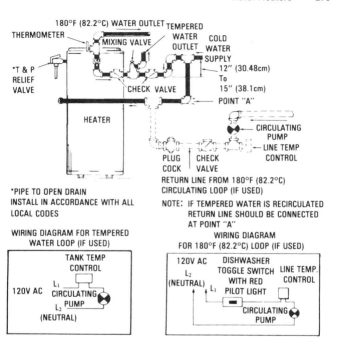

Fig. 16-7. Water line connection to a two-temperature, one-heater, high-temperature storage model, with or without recirculation. *(Courtesy A.O. Smith Corporation, Consumer Products Division)*

A diaphragm expansion tank is provided on top of the heater to handle the expansion of heat transfer fluid in a closed-loop circulating line.

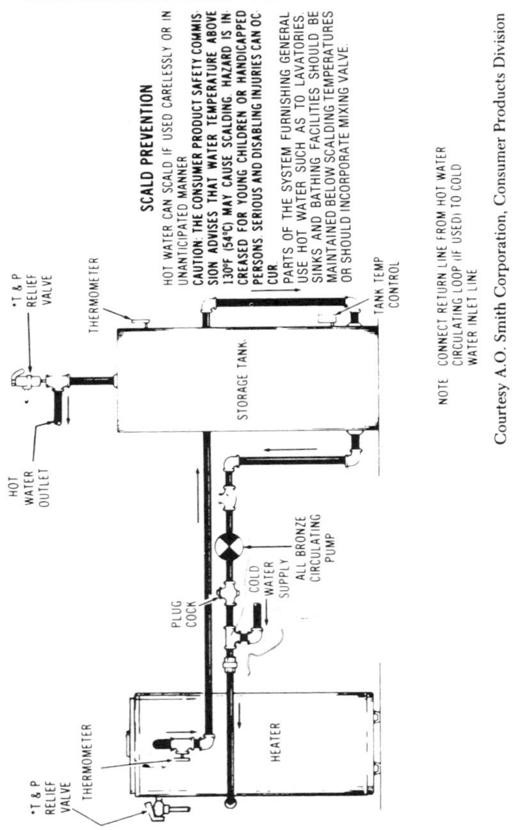

SCALD PREVENTION

HOT WATER CAN SCALD IF USED CARELESSLY OR IN UNANTICIPATED MANNER

CAUTION: THE CONSUMER PRODUCT SAFETY COMMISSION ADVISES THAT WATER TEMPERATURE ABOVE 130F (54°C) MAY CAUSE SCALDING. HAZARD IS INCREASED FOR YOUNG CHILDREN OR HANDICAPPED PERSONS. SERIOUS AND DISABLING INJURIES CAN OCCUR.

PARTS OF THE SYSTEM FURNISHING GENERAL USE HOT WATER SUCH AS TO LAVATORIES SINKS AND BATHING FACILITIES SHOULD BE MAINTAINED BELOW SCALDING TEMPERATURES OR SHOULD INCORPORATE MIXING VALVE

*T & P RELIEF VALVE

THERMOMETER

TANK TEMP CONTROL

STORAGE TANK

HOT WATER OUTLET

NOTE: CONNECT RETURN LINE FROM HOT WATER CIRCULATING LOOP (IF USED) TO COLD WATER INLET LINE

PLUG COCK

COLD WATER SUPPLY

ALL BRONZE CIRCULATING PUMP

THERMOMETER

HEATER

*T & P RELIEF VALVE

Courtesy A.O. Smith Corporation, Consumer Products Division

Fig. 16-8. Water line connections to a one-temperature, one-heater model with vertical storage tank forced circulation, with or without building recirculation. *(Courtesy A.O. Smith Corporation, Consumer Products Division)*

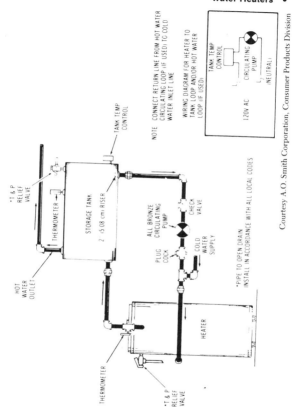

Courtesy A.O. Smith Corporation, Consumer Products Division

Fig. 16-9. Water line connections to a one-temperature, one-heater model with horizontal storage tank forced circulation, with or without building recirculation. *(Courtesy A.O. Smith Corporation, Consumer Products Division)*

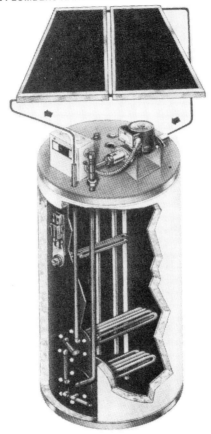

Fig. 16-10. The Conservationist Solar System Water Heater. *(Courtesy A.O. Smith Corporation, Consumer Products Division)*

Two high-density magnesium anodes protect the tank against corrosion.

A three-inch double efficiency blanket of high-density insulation surrounds the tank to keep in more heat.

The tank is isolated from the jacket to prevent conduction heat loss.

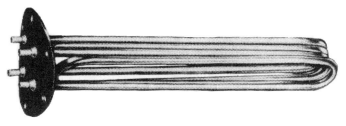

Fig. 16-11. Two elements in one unit—either or both may be operated to control how fast the heater can produce hot water.

Exclusive Corona™ Heat Exchanger

The heat exchanger is immersed in the tank to ensure direct transfer of the heat. Ordinary exchangers are less efficient with a wrap-around-tank method. The Corona has a double wall of copper for safety and is electrically isolated from the tank and external piping for positive protection against corrosion.

The Corona heat exchanger is used in Conservationist solar water heaters — models Sun-82, 100 and 120 gallon, or 378.5 and 454.25 liter sizes.

Fig. 16-12. Single element "life belt" type heating element.

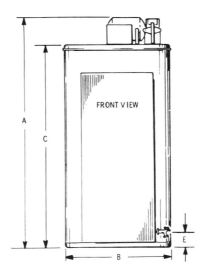

Fig. 16-13. Solar system water heater.

Table 16-2. Solar System Water Heater Specifications

Model No.	A		B		C		E	
	Inches	mm	Inches	mm	Inches	mm	Inches	mm
Sun-82	56	1422	28	711	48	1219	4½	108
Sun-100	65⅞	1673	28	711	57⅞	1470	4½	108
Sun-120	69	1753	30	762	61	1549	5¾	146

Model No.	Capacity		Approximate Shipping Weight	
	U.S. Gallons	Liters	Pounds	Kilograms
Sun-82	82	310.3	235	106.5
Sun-100	100	378.5	250	113.4
Sun-120	120	454.25	340	154.2

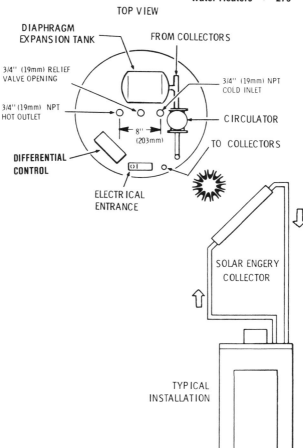

TOP VIEW

DIAPHRAGM
EXPANSION TANK

FROM COLLECTORS

3/4" (19mm) RELIEF
VALVE OPENING

3/4" (19mm) NPT
COLD INLET

3/4" (19mm) NPT
HOT OUTLET

CIRCULATOR

8"
(203mm)

TO COLLECTORS

DIFFERENTIAL
CONTROL

ELECTRICAL
ENTRANCE

SOLAR ENGERY
COLLECTOR

TYPICAL
INSTALLATION

Fig. 16-14. Using solar energy to heat water.

The Phoenix™ screw-in immersion element has two-way protection: A sheathing of iron-base superalloy provides excellent protection against burn-out, oxidation, and scaling. The ceramic terminal block will not melt like ordinary plastic blocks. The Phoenix elements provide back-up heating, as needed.

17
Water Coolers

Water coolers are used in schools, industry, and hospitals.
They require both plumbing and electrical attention. Various
installations are covered in this chapter.

The numbers below refer to Fig. 17-1.

1. The Dial-A-Drink Bubbler™ assures a smooth, even flow
   of water under pressures from 20 to 125 psi (138 to 862
   kPa).
2. The stainless steel top of polished 18-8 stainless resists rust,
   corrosion, and stains. An anti-splash ridge and integral
   drain direct and dispose of water.
3. The red brass cooling tank offers maximum cooling efficiency and reduces starts of the compressor. The 85-15
   red brass storage tank (vented) has an internal heat
   transfer surface and external refrigeration coil bonded to
   the tank by immersion in pure molten tin.
4. The copper cooling coils around the storage tank ensure
   maximum cooling efficiency. Double-wall separation of
   the refrigerant and drinking water conforms to all sanitary
   codes.
5. An insulating jacket of expanded polystyrene foam main-
   tains cold water temperature on all models.
6. An adjustable thermostat is tamper-proof. A remote-
   sensing bulb is located in the cooling tank and provides
   accurate control of cold water temperature.
7. The cost cutting pre-cooler (on larger capacity models)
   nearly doubles the capacity without extra operating cost
   by cooling incoming water with cold waste water.

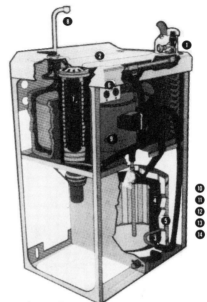

Fig. 17-1. Wall-hung water cooler.

8. Hot water availability on Hot 'N Cold models: The hot tank heats and serves up to 45 cups of piping hot water per hour. The hot water system is atmospherically vented and fiberglass insulated.
9. The refrigeration system is maintenance free. The compressor and motor are hermetically sealed, lubricated for life, and leakproof, if properly installed.
10. The durable cabinet finish includes vinyl laminated-on steel on the front and side panels, which provides a scuff-resistant finish.

Dimensional Drawing
Dimensions are in inches (millimeters)

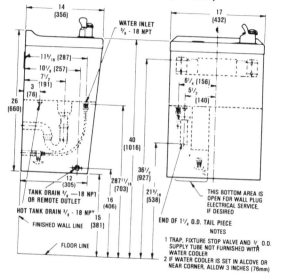

WATER INLET
³/₈ - 18 NPT

14 (356)
11⁵/₁₆ (287)
10¹/₈ (257)
7¹/₂ (191)
3 (76)
26 (660)
40 (1016)
36¹/₂ (927)
287¹¹/₁₆ (703)
16 (406)
21³/₈ (538)
15 (381)
12 (305)

TANK DRAIN ³/₈ - 18 NPT
OR REMOTE OUTLET

HOT TANK DRAIN ³/₈ - 18 NPT
FINISHED WALL LINE

FLOOR LINE

17 (432)
6¹/₄ (156)
5¹/₂ (140)

THIS BOTTOM AREA IS
OPEN FOR WALL PLUG
ELECTRICAL SERVICE,
IF DESIRED

END OF 1¹/₄ O.D. TAIL PIECE

NOTES

1 TRAP, FIXTURE STOP VALVE AND ³/₈ O.D.
SUPPLY TUBE NOT FURNISHED WITH
WATER COOLER
2 IF WATER COOLER IS SET IN ALCOVE OR
NEAR CORNER, ALLOW 3 INCHES (76mm)

FOR VENTILATION

Number of People Served

MODEL	GHP of 50°	Offices	Light Industry
ODP16M	15.7	188	100
ODP13M	13.0	156	91
ODP13M60	13.0	156	91
ODP7M	7.0	84	49
ODP7MH	7.0	84	49
ODP5M	5.0	60	35
ODPM	Non Refrigerated Fountain		
ODP15MW	Water cooled		

Fig. 17-2. Dimensions for mounting water cooler.

11. ARI certified performance means the cooling capacity of a water cooler is rated and certified in accordance with Air Conditioning and Refrigeration Institute (ARI) Standard 1010-73 (A.N.S.I. Standard A112.-11.1-1973): Gallons per hour of 50°F (10°C) drinking water with inlet temperature of 80°F (27°C), and room temperature of 90°F (32°C).

12. Most semirecessed and simulated semirecessed models have removable front and side panels, which provide extra work space for plumbing and electrical installations, servicing, and routine maintenance.

13. Each water cooler is performance tested.

Installation instructions are shown in Fig. 17-2:

Rough in above water cooler at:

Waste: 22½″ or 572 mm, 5½″ (140 mm) left of center line.

Water: 17½″ or 445 mm, 6½″ (165 mm) left of center line.

Water supply pipe: ½″ N.P.S. or 13 mm — reduced to ³/₈″ O.D. (10 mm). Waste piping is 1¼″ I.D. or 32 mm.

PART III

General Reference Information

18
Abbreviations, Definitions, and Symbols

Abbreviations

A.G.A. — American Gas Association
A.S.A. — American Standards Association
A.S.H.V.E. — American Society of Heating & Ventilation
F. & D. — Faced and drilled
I.B.B.M. — Iron body bronze or brass mounted
M.S.S. — Manufacturer Standardization Society of Valve &
    Fittings Industry
N.P.S. — Nominal Pipe Size
O.S. & Y. — Outside screw & yoke
L.I.A. — Lead Industries Association
R.N.P.T. — Right-hand national pipe thread
N.D.T.S. — Not drawn to scale
A.S.T.M. — American Society for Testing Material
C.A.B.R.A. — Copper & Brass Research Association
C.I.S.P.I. — Cast Iron Soil Pipe Institute
B.M. — Bench mark

Definitions

Following are definitions of terms and quantities, plus facts and tips for plumbers. Also see the metric section.

1 lb. of air pressure elevates water approximately 2.31 feet under atmospheric conditions of 14.72 psi.

2.3 feet of water = 1 psi.

1 foot of water = 0.434 psi.

1.728 cubic inches = 1 cubic foot.

231 cubic inches in one U.S. gallon.

One cubic foot of water at 39°F weighs 62.48 lbs.

One U.S. gallon of cold water weighs 8.33 lbs.

One cubic foot of water contains 7.48 gallons.

SI

In SI (metric system), 1 kilopascal (kPa) of air pressure elevates water approximately 10.2 cm under atmospheric conditions of 101 kPa.

10.2 cm of water = 1 kPa.

51 cm of water = 5 kPa.

1 meter of water = 9.8 kPa.

10,000 square centimeters = 1 square meter.

1 cubic meter = 1,000,000 cubic centimeters (cm^3) or 1000 cubic decimeters (dm^3).

1 liter of cold water at 4°C weighs 1 kilogram.

Air Pressure

One cubic inch of mercury weighs 0.49 lbs. Therefore, a 10″ column of mercury would be 10 × 0.49 or 4.9 psi. Generally speaking, 2″ of mercury = 1 lb. pressure.

One cubic foot of air weighs 1.2 ounces or 0.075 lbs.

Atmospheric pressure of 14.7 psi will balance or support a column of mercury 29.92 inches high.

Absolute Zero
Absolute zero is –459.69°F.

Absolute zero is –273.16°C.

In SI, 1 centimer of mercury at 0°C = 1.3332239 kPa pressure. Therefore, a 24 cm column of mercury would be 24 × 1.3332239 or, rounded, 32 kPa. Generally speaking, 6 cm of mercury = 8 kPa.

1 cubic meter of air weighs 1.214 kilograms.

Atmospheric pressure of 101.3 kPa will balance or support a column of mercury 76 cm high.

Minimum Grade Fall
The absolute minimum fall or grade for foundation or subsoil drainage lines is 1″ (2.5 cm) in 20′ (6.1 m).

Cylinder Pressure
Cylinders are charged with oxygen at a pressure of 2000 psi (13789 kPa) at 70°F (21°C).

Welding Flame
The temperature of an oxygen-acetylene flame is estimated to be over 6000°F (3316°C).

Gaskets
Suitable gasket material for a flange union should be made of:

• Cold water piping — sheet rubber or asbestos° sheet packing.

• Hot water lines — rubber or asbestos° composition.

°Asbestos is not permitted in any form in some local codes.

- Gas piping — <u>leather</u> or asbestos° composition.
- Oil lines — metallic or, where permitted, asbestos° composition.
- Gasoline conduction — <u>metallic.</u>

By applying graphite to one side of a gasket, removal at a later date is made much easier.

Forming an Angle Using a Folding Rule

You can form commonly used angles by the use of a folding rule using its first four sections. This will enable you to determine what fittings will work in offsetting situations, or determine the proper direction to turn a pipe when it is necessary to come out of a wall on an angle. To form one of these angles or bends, see the examples that follow.

$22\frac{1}{2}°$ or $\frac{1}{16}$ bend: take the tip of the rule and touch $23\frac{3}{4}''$; straighten out the rule at the second joint for the angle.

$11\frac{1}{4}°$ or $\frac{1}{32}$ bend: take the tip of the rule and touch $23\frac{15}{16}''$. Follow as previously mentioned.

$30°$ angle: touch the tip of the rule to $14\frac{13}{16}''$. Follow as before.

$45°$ or $\frac{1}{8}$ bend: touch the tip of the rule to $23''$. Follow as before.

$60°$ or $\frac{1}{6}$ bend: touch $22\frac{1}{4}''$ with the tip. Follow as before.

$72°$ or $\frac{1}{5}$ bend: touch the tip of the rule to $21\frac{5}{8}''$. Follow as before.

$90°$ angle: touch $20\frac{1}{4}''$ with the tip. Follow as before.

°Asbestos is not permitted in any form in some local codes.

American Standard Threads

Pipe Size	Threads Per Inch	Threads Per Centimeter
⅛" or 0.3175 cm	27	10½
¼" or 0.635 cm	18	7
⅜" or 0.9525 cm	18	7
½" or 1.27 cm	14	5½
¾" or 1.905 cm	14	5½
1" or 2.54 cm	11½	4½
1¼" or 3.175 cm	11½	4½
1½" or 3.81 cm	11½	4½
2" or 5.08 cm	11½	4½
2½" through 4" or 6.35 through 10.16 cm	8	3⅛

Machine Screw Bolt Information (NC)

Size	Diameter in Inches or Millimeters To nearest 1/64" (½ mm)
1	5/64" (2 mm)
2	5/64" (2 mm)
3	7/64" (2½ mm)
4	7/64" (3 mm)
5	⅛" (3 mm)
6	9/64" (3½ mm)
8	11/64" (4 mm)
10	13/64" (5 mm)
12	13/64" (5½ mm)
¼"	¼" (6 mm)

Note: Above No. 12, machine screw sizes are designated by actual diameter.

Standard Wood Screw Information

Size Number	Decimal	Diameter	
		Inches	mm*
0	0.064	$\frac{1}{16}$	1½ mm
1-2	0.077, 0.090	$\frac{3}{32}$	2 mm
3-4-5	0.103, 0.116, 0.129	$\frac{1}{8}$	2½ to 3 mm
6-7-8	0.142, 0.155, 0.168	$\frac{5}{32}$	3½ to 4 mm
9-10-11-12	0.181, 0.194, 0.207, 0.220	$\frac{3}{16}$	4½ to 5½ mm
14-16	0.246, 0.272	$\frac{1}{4}$	6 to 7 mm
18-20	0.298, 0.324	$\frac{5}{16}$	7½ to 8 mm
24	0.376	$\frac{3}{8}$	9½ mm

Flathead screws are measured by overall length; roundhead screws from base of head to end.

*to nearest ½ mm

Hanger Rod Sizes

Iron Pipe Size	Rod Size
$\frac{1}{8}$"-½" (3-13 mm)	¼" (6 mm)
¾"-2" (19-51 mm)	$\frac{3}{8}$" (10 mm)
2½" and 3" (64 and 76 mm)	½" (13 mm)
4" and 5" (10 and 13 cm)	$\frac{5}{8}$" (16 mm)
6" (15 cm)	¾" (19 mm)
8", 10", 12" (20, 25, 30 cm)	$\frac{7}{8}$" (22 mm)
14" and 16" (36 and 41 cm)	1" (25 mm)

Hanger rod is threaded with (NC) national coarse bolt dies. A threading die marked, $\frac{5}{8}$"-11 NC, would be used for $\frac{5}{8}$" diameter rod, the thread classified as national coarse with 11 threads per inch.

Boiling Points of Water at
Various Pressures Above Atmospheric

Atmospheric, or 0 gage pressure: Boiling point 212°F (100°C).

Gage Pressure psi or kPa	Boiling Point
1-6.89	216° F (102.2° C)
4-27.58	225° F (107.2° C)
15-103.43	250° F (121.1° C)
25-172.36	267° F (130.5° C)
30-206.84	274° F (134.4° C)
45-310.26	293° F (145.0° C)
50-344.73	297° F (147.2° C)
65-448.13	312° F (155.5° C)
75-517.1	320° F (160.0° C)
90-620.52	335° F (168.3° C)
100-689.47	338° F (170.0° C)
125-861.83	353° F (178.3° C)
150-1034.2	366° F (185.5° C)

Melting Points

Lead melts at 622°F (328°C).
Tin melts at 449°F (231.78°C).
50-50 solder begins to melt at 362°F (183.3°C).
Zinc melts at 790°F (421°C).
Pure iron melts at 2,730°F (1,499°C).
Steel melts from 2,400 to 2,700°F (1,315.5°C to 1,482.2°C).

Symbols for Plumbing Fixtures

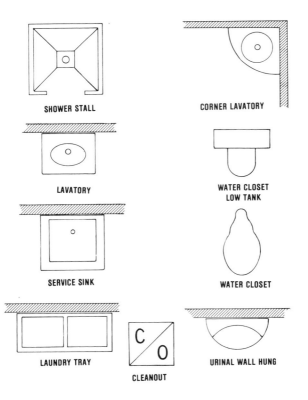

SHOWER STALL

CORNER LAVATORY

LAVATORY

WATER CLOSET
LOW TANK

SERVICE SINK

WATER CLOSET

LAUNDRY TRAY

C
O

CLEANOUT

URINAL WALL HUNG

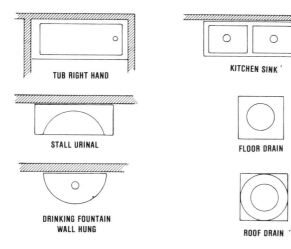

TUB RIGHT HAND

KITCHEN SINK

STALL URINAL

FLOOR DRAIN

DRINKING FOUNTAIN
WALL HUNG

ROOF DRAIN

19
Formulas

Table of Squares

$11^2 = 121$	$19^2 = 361$
$12^2 = 144$	$20^2 = 400$
$13^2 = 169$	$21^2 = 441$
$14^2 = 196$	$22^2 = 484$
$15^2 = 225$	$23^2 = 529$
$16^2 = 256$	$24^2 = 576$
$17^2 = 289$	$25^2 = 625$
$18^2 = 324$	$26^2 = 676$

Finding Square Roots

Step 1

Pointing off the number to be worked in preparation to a solution is the first and most important step.

Whether working with a decimal or whole number, always point off in two's, starting from left to right. When a decimal is present and the numbers are odd, a zero (0) must be added:

$\sqrt{25.325}$, should be $\sqrt{25.3250}$.

Examples:

$$\sqrt{2'59} \quad \sqrt{25.'32} \quad \sqrt{25.'32'50}$$

Step 2

Beginning with the first number or set of numbers on the left, we begin our problem; in the three examples above, the numbers would be (2) (25) (25) respectively.

We must find the nearest square (or number multiplied by itself) to this number but not to total higher than the number. In the examples above, squares would be: (1), (5), (5) respectively.

Step 3

```
        5
      × 5
      ———
       25              [1]  [4] [6]
                        5.  0  3
      100            √ 25.'32'50'
  [6] + 3        [1]   25
     ————        ———  ————————
     1003        [3]  50|32 [2]
                       3250 [5]
     1003             −3009
      × 3             ———————
     ————              241 [7]
     3009
```

1. $5 \times 5 = 25$ (place as shown)
2. Bring down the 32.
3. Try dividing 32 by 50 (the 5 + a trial 0).
4. 32 is less than 50 so place 0 above the 32.
5. Bring down the 50 that is alongside the 32 and place as shown.
6. Multiply the 50 above by 2 to obtain 100. Take this 100 and add a trial number (in this case a 3) and multiply the 1003 by 3. The 3 is the number then used to make sure the 3250 is not exceeded when 3 is multiplied by 1003. If this exceeded the 3250, then you'd have to drop back and try a 2, etc., until you you get a result less than that desired.
7. Subtract 3009 from 3250 and obtain 241.
8. The 241 may be left as a remainder or you may continue to obtain 5.0323951. However, two places beyond the decimal point is usually sufficient for this type of work.

This can be continued until the square root is 5.0323951. However, today very inexpensive calculators save time and improve accuracy. Just enter 25.325 and then press the [√⁻] button to instantly get the answer to 7 decimal places.

Problem Proved:

$$
\begin{array}{r}
5.03 \\
\times\ 5.03 \\
\hline
1509 \\
25150 \\
\hline
25.3009 \\
+\ 241 \quad = R \\
\hline
25.3250
\end{array}
$$

Finding Diagonal of a Square

The diagonal of a square equals the square root of the sum of the squares of the two sides (Fig. 19-1).

Example: Find the diagonal of a square when the area is 8100 square inches: $(90^2 = 8100)$. Remember, area = length × width.

[1] $8100 \times 2 = 16{,}200$

[2] $16{,}200$ sq. in. = twice the area

[3] $\sqrt{16{,}200} = 127.27922$ or

[4] rounded to $127\frac{1}{4}''$ (Fig. 19-1).

Note: Figures used in the example could also be centimeters or meters, etc.

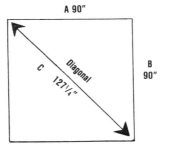

Fig. 19-1. A square produces two right angles when bisected or cut in half.

Right Triangle

C = Hypotenuse
A = Altitude
B = Base

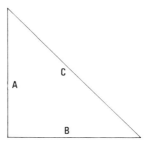

Formula: $C^2 = A^2 + B$

Note: The square of the hypotenuse equals the sum of the squares of the other two sides (Pythagorean Theorem).

$3 \times 3 = 9$
$4 \times 4 = \underline{16}$
25

$$\begin{array}{r} 5 \\ \times\ 5 \\ \hline 25 \end{array}$$

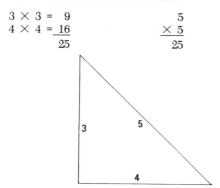

Note: To square your work while fabricating or forming a square, use 3 – 4 – 5 method. You may use 3 ft., 4 ft., 5 ft., or you may use 3″, 4″, 5″; to increase the sides of the square, merely double or triple each number; for example (6 – 8 – 10) (9 – 12 – 15) and so on.

Problems:

Running pipe or tubing in parallel runs using offsets and maintaining a uniform spread throughout the run.

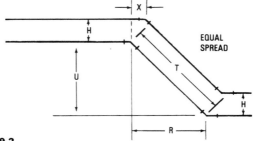

Fig. 19-2.

Formulas for Fig. 19-2

$22^1/_2°$ ells or $^1/_8$ bends $\quad$ 45° ells or $^1/_8$ bends $\quad$ 60° ells or $^1/_6$ bends

T = U × 2.61	T = U × 1.41	T = U × 1.15
or	or	or
T = R × 1.08	T = R × 1.41	T = R × 2
U = R × 0.41	U = R	U = R × 1.73
R = U × 2.41	R = U	R = U × 0.58
X = H × 0.20	X = H × 0.41	X = H × 0.58

Formulas for Fig. 19-3

X = H × 0.41 $\qquad$ C = B + (H × 0.41 × 2) or

B = A + (H × 0.41 × 2) or $\quad$ B = C – (H × 0.41 × 2)

A = B – (H × 0.41 × 2)

Figure 19-4 illustrates: 45° Y and $^1/_6$ bend in an offset:
Formula for above:
S = U
R = S × 1.41
T = S × 2, or U × 2, or
$\quad\quad$ R × 1.41
U = Vertical rise
S = Horizontal spread
R = Advance or setback
T = Travel or pipe to be cut
T will be C-C-measurement

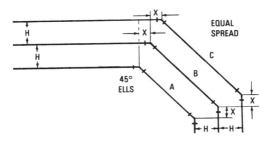

Fig. 19-3.

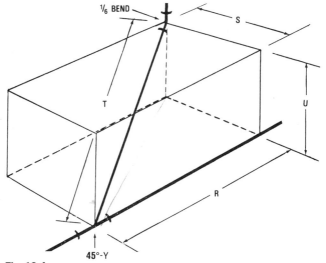

Fig. 19-4.

Determining Capacity of a Tank in Gallons

Example: Tank is 4′ 0″ in diameter and 10′ 0″ long.

(A) Find the area of the circle: diameter squared × 0.7854:

Thus, 4′ × 4′ × 0.7854:
Thus, 4′ × 4′ × 0.7854 = 12.566 square feet.

(B) Find the cubic contents: area of a circle × length:

12.566 sq. ft. × 10′ = 125.66 cu. feet

(C) Find the number of gallons: multiply the cubic contents by 7.48 (the number of gallons in one cubic foot):

125.66 cu. ft. × 7.48 gals. = 939.936, or 940 gals.

Determining Capacity of a Tank in Liters

Example: Tank is 1.22 meters in diameter and 3.05 meters long —

(A) Find the area of the circle: diameter squared × 0.7854:

Thus, 1.22 × 1.22 × 0.7854:
Thus, 1.22 × 1.22 × 0.7854 = 1.169 sq. meters

(B) Find the cubic contents: area of a circle × length:

1.169 sq. meters × 3.05 meters = 3.565 cu. meters

(C) Find the number of liters: multiply the cubic contents by 1000 (the number of liters in one cubic meter):

3.565 cu. meters × 1000 = 3565 liters.

Area of triangle: $A = \frac{1}{2} BH$
Area of circle: $= \pi R^2$
    or $0.7854 D^2$

Note: $\pi = 3.1416$

To find the area of a circle when the circumference is known:

$$A = \frac{C^2}{12.57}$$

If a circle is 45ˣ (x may be inches, feet, meters, centimeters, or or millimeters)

$$A = \frac{45 \times 45}{12.56}$$

Answer: A = 161.09ˣ

To find the head when pressure is given, divide the pressure by 0.433, or multiply the pressure by 2.309.

Formula: psi = H × 0.433
H = psi × 2.309

Rolling Offsets

Formula using 45° fitting:

$X = \sqrt{S^2 + U^2}$
$T = X \times 1.41$
$R = X$

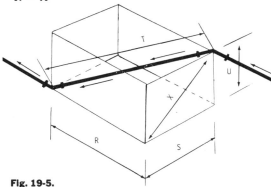

Fig. 19-5.

Formula using 60° fittings:

$X = \sqrt{S^2 + U^2}$
$T = X \times 1.15$
$R = X \times 0.58$

Formula using 22½° fittings:

$X = \sqrt{S^2 + U^2}$
$T = X \times 2.61$
$R = X \times 2.41$

Example:
Using 45° fittings, assume U is 25 and S is 30, so U must be squared:

Step 1
$25 \times 25 = 625$ (U squared)

Step 2
S must be squared: or $30 \times 30 = 900$

Step 3
$$
\begin{aligned}
\text{Add } U^2 &= \quad 625 \\
S^2 &= +\ \ 900 \\
\hline
&\quad 1525
\end{aligned}
$$

Find the square root of 1525 (calculator solution):

$$
\begin{array}{c}
39.05 \\
\sqrt{1525.00} \\
= \text{sq. root R4}
\end{array}
$$

Proved:
$$
\begin{aligned}
39.05 \\
\times\ 39.05 \\
\hline
1524.9025
\end{aligned}
$$

Step 4

X = 39
T = Travel or pipe to be cut

```
   1.41
 ×  39
  1269
   423
 54.99
```

Answer: 55 C-C

Fraction-Decimal Equivalents

Fraction — Change to decimal when multiplying:

Fraction	Decimal
1/16″	0.06
1/8″	0.13
3/16″	0.19
1/4″	0.25
5/16″	0.31
3/8″	0.38
7/16″	0.44
1/2″	0.50
9/16″	0.56
5/8″	0.63
11/16″	0.69
3/4″	0.75
13/16″	0.81
7/8″	0.88
15/16″	0.94

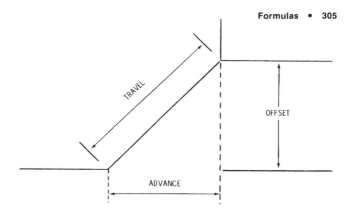

Constants for Calculating Offset Measurements

Degree of Fitting	Known Constant Factor
$5\frac{5}{8}°$	Offset × 10.20 = Travel
$5\frac{5}{8}°$	Offset × 10.152 = Advance
$11\frac{1}{4}°$	Offset × 5.126 = Travel
$11\frac{1}{4}°$	Offset × 5.027 = Advance
$22\frac{1}{2}°$	Offset × 2.613 = Travel
$22\frac{1}{2}°$	Offset × 2.414 = Advance
30°	Offset × 2.000 = Travel
30°	Offset × 1.732 = Advance
45°	Offset × 1.414 = Travel
45°	Offset × 1.00 = Advance
60°	Offset × 1.155 = Travel
60°	Offset × 0.577 = Advance
$67\frac{1}{2}°$	Offset × 1.083 = Travel
$67\frac{1}{2}°$	Offset × 0.414 = Advance
72°	Offset × 1.052 = Travel
72°	Offset × 0.325 = Advance

Note: When above constants are used to calculate an offset measurement, the given offset and solution are expressed in center-to-center measurements.

20
Metric Information Helpful
to the Piping Industry

Metric Abbreviations

ANMC . . . . American National Metric Council
ANSI . . . . American National Standards Institute
NBS . . . . National Bureau of Standards
SI . . . . International System of Units
m . . . . meter
km . . . . kilometer
cm . . . . centimeter
mm . . . . millimeter
kg . . . . kilogram
g . . . . gram
mg . . . . milligram
l . . . . liter
ml . . . . milliliter
Nm . . . . newton-meter
°C . . . . degrees celsius
kPa . . . . kilopascal
J . . . . joule
inHg . . . . inches of mercury
Pa . . . . pascal

Information

Length

Metric	English
1 meter = 39.37 inches	
1 meter = 1000 millimeters	

Length (*continued*)

1 meter = 100 centimeters
1 meter = 10 decimeters
1 kilometer = 0.625 miles
1.609 kilometers = 1 mile
25.4 millimeters = 1 inch
2.54 centimeters = 1 inch
304.8 millimeters = 1 foot
1 millimeter = 0.03937 inches
1 centimeter = 0.3937 inches
1 decimeter = 3.937 inches

Volume — Liquid

Metric	English
3.7854 liters	= 1 gallon
0.946 liters	= 1 quart
0.473 liters	= 1 pint
1 liter	= .264 gals. or 1.05668 qts.
1 liter	= 33.814 ounces
29.576 milliliters	= 1 fluid ounce
236.584 milliliters	= 1 cup

Note: 1 liter contains 1000 milliliters

Inches to Millimeters		Feet to Millimeters	
English (inches)	Metric (mm)	English (feet)	Metric (mm)
1	25.4	2	609.6
2	50.8	3	914.4
3	76.2	4	1219.2
4	101.6	5	1524.0
5	127	6	1828.8
6	152.4	7	2133.6
7	177.8	8	2438.4
8	203.2	9	2743.2
9	228.6	10	3048.0
10	254	20	6096.0
11	279.4		
12	304.8		

Note: Round off to nearest millimeter. Thus,
4″ = 102 mm
3″ = 76 mm
2″ = 51 mm, and so on.

Parts of an Inch to Millimeters

English (parts of an inch)	Metric (mm)	English (parts of an inch)	Metric (mm)
1/32	0.79375 (0.80)	9/16	14.2875 (14.3)
1/16	1.5875 (1.6)	5/8	15.8750 (15.9)
1/8	3.175 (3.2)	11/16	17.4625 (17.5)
3/16	4.7625 (4.8)	3/4	19.0500 (19.1)
1/4	6.35 (6.4)	13/16	20.6375 (20.6)
5/16	7.9375 (7.9)	7/8	22.2250 (22.2)
3/8	0.5250 (9.5)	15/16	23.8175 (23.8)
7/16	11.1125 (11.1)	1	25.4000 (25.4)
1/2	12.7000 (12.7)		

Note: In most cases it is best to round off to the nearest millimeter. Thus,
17.4625 would be: 17 mm
20.6375 would be: 21 mm

Inches and Parts of an Inch to Centimeters

Inches (parts of an inch)	Metric (cm)	Inches (parts of an inch)	Metric (cm)
1/16	0.15875	2 1/2	6.35
1/8	0.3175	3	7.62
1/4	0.635	4	10.16
3/8	0.9525	5	12.70
1/2	1.27	6	15.24
5/8	1.5875	7	17.78
3/4	1.905	8	20.32
7/8	2.2225	9	22.86
1	2.54	10	25.40
1 1/4	3.175	11	27.94
1 1/2	3.81	12	30.48
2	5.08		

Inches and Feet to Meters

Inches	Meters	Feet	Meters
1	0.0254	1¼	0.381
2	0.0508	1½	0.4572
3	0.0762	2	0.6096
4	0.1016	2½	0.762
5	0.127	3	0.9144
6	0.1524	4	1.2192
7	0.1778	5	1.524
8	0.2032	6	1.8288
9	0.2286	10	3.048
10	0.254	25	7.62
11	0.2794	50	15.24
12	0.3048	100	30.48

Temperature

Fahrenheit		Celsius
212°	← Temperature of Boiling Water →	100°
176°		80°
140°		60°
122°		50°
104°		40°
98.6°	← Temperature of Human Body →	37°
95°		35°
86°		30°
77°		25°
68°		20°
50°		10°
32°	← Temperature of Melting Ice →	0°
−4°		−20°
−40°	← Temperature Equal →	−40°
−459.69°	← Absolute Zero →	−273.16°

The following formulas may be used for converting temperatures given on one scale to that of the other.

When using a calculator it is easier to compute:

F = 1.8C + 32
   or 1.8 $\times$ C + 32

C = F - 32 $\div$ 1.8
   or F minus 32 divided by 1.8

Pressure

Kilopascal (kPa) is the unit recommended for fluid pressure for almost all fields of use, such as barometric pressure, gas pressure, tire pressure, and water pressure.

Atmospheric pressure is: 101 kPa metric, and 14.7 psi English; 6.894757 kPa = 1 psi.

To find head pressure in decimeters when pressure is given in kilopascal (kPa), divide pressure by 0.9794.

To find pressure in kPa of a column of water given in decimeters, multiply decimeters by by 0.9794.

To find head in meters when pressure is given in kilopascals, divide pressure by 9.794.

To find pressure in kPa of a column of water given in meters, multiply meters by 9.794.

Example:

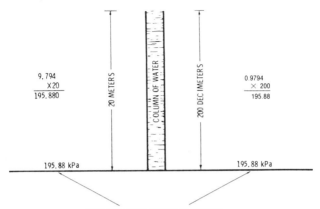

PRESSURE AT BASE OF WATER COLUMN

Miscellaneous Metric Information

The amount of heat required to change ice to liquid water is 144 Btu's per lb. (335 joules per kilogram).

Generally speaking, 6 cm of mercury = 8 kPa pressure.

One cubic meter of air weighs 1.214 kilograms (kg).

Atmospheric pressure of 101.3 kPa will balance or support a column of mercury 76 cm high.

When you know inches of mercury (inHg), multiply this quantity by 3.386389 to find the number of kilopascals (kPa).

Column of water: 9.794 kPa per meter (see Example above), or 0.2476985 kPa per inch.

1 ml of water has a mass of 1 gram.

1 foot-lb. = 1.3558 newton-meter (Nm) (bending moment of torque).

Note: 1 British thermal unit (mean) = 1.05587 joule.
   1 Btu (international table) = 1.055056 joule.
Example: 144 Btu's = 152 joule (J)
Use conversion factor: 1.056
                       or 1 Btu = 1.056 joule

1 kPa of air pressure elevates water approximately 10.2 cm under atmospheric conditions of 101 kPa.

10.2 cm of water = 1 kPa.
51 cm of water = 5 kilopascals.
10,000 sq. cm = 1 sq. meter.
1 cubic meter = 1,000,000 cu. cm or 1000 cu. dm.
1 liter of cold water at 4°C weighs 1 kilogram.

1 centimeter column of mercury at 0°C = 1.3332239 kPa pressure. Therefore, a 24-cm column of mercury would be 24 × 1.3332239 or rounded 32 kPa. Generally speaking, 6 centimeters (cm) of mercury = 8 kilopascals (kPa).

Note: To determine degrees Fahrenheit when Celsius is given: °F = 1.8C + 32 or 1.8 × C + 32. To determine: degrees Celsius when Fahrenheit is given: °C = F − 32 ÷ 1.8 or F − 32 then ÷ 1.8.

(Acceptable Forms)	(Not Acceptable)
12 to 20°C	20° C
12°C to 20°	12° to 20° C

Square and Cube Measures

2.59 km²	1 square mile
0.093 m²	1 square foot
6.451 cm²	1 square inch
0.765 m³	1 cubic yard
0.028316 m³	1 cubic foot
16.387 cm³	1 cubic inch
1 cubic meter (m³)	35.3146 cubic feet
929.03 sq. centimeters (cm²)	1 square foot

Note:

10,000 sq. centimeters (cm²)	1 m²

1 cubic meter (m^3) . . . . 1,000,000 cm^3 or 1000 dm^3
10.2 cm of water . . . . 1 kPa of pressure
51 cm of water . . . . .5 kPa
1 meter of water . . . . 9.8 kPa
1 cubic foot contains . . . . 28,316.846522 cm^3

Weight (Mass)

1 kilogram = 2.204623 lbs.
453.592 grams = 1 kilogram
1 gram = 0.035 ounces
28.349 grams = 1 ounce
28,349 milligrams = 1 ounce
1 gram = 1000 milligrams
1 kilogram = 1,000,000 milligrams
1 kilogram also = 1000 grams
0.02831 kilograms = 1 ounce

Note:

1 lb. = 453,592.37 milligrams
1 lb. = 453.59237 grams
1 lb. = 0.453592 kilograms
1 metric ton weighs 1000 kilograms or 2204.623 lbs.

21
Knots Commonly Used

Bowline

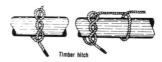

Timber hitch

Barrel hitch

Square knot

Bowline on a bight

Half hitch and two half hitches

22
Typical Hoisting Signals

Lower load

Boom down

Stop

Travel forward

Swing or house

Travel backward

Boom up

Move slowly

Emergency stop

Index

A

B

C

MONROE CONSTRUCTION, INC.
133 Central Street
WARWICK, RHODE ISLAND 02886-1284

MONROE CONSTRUCTION, INC.
133 Central Street
WARWICK, RHODE ISLAND 02886-1284

MONROE CONSTRUCTION, INC.
133 Central Street
WARWICK, RHODE ISLAND 02886-1284